Fundamental
Forces of Nature

The Story of Gauge Fields

Fundamental Forces of Nature

Nature

The Story of Gauge Fields

Kerson Huang

Massachusetts Institute of Technology, USA

World Scientific

NEW JERSEY · LONDON · SINGAPORE · BEIJING · SHANGHAI · HONG KONG · TAIPEI · CHENNAI

Published by

World Scientific Publishing Co. Pte. Ltd.
5 Toh Tuck Link, Singapore 596224
USA office: 27 Warren Street, Suite 401-402, Hackensack, NJ 07601
UK office: 57 Shelton Street, Covent Garden, London WC2H 9HE

British Library Cataloguing-in-Publication Data
A catalogue record for this book is available from the British Library.

First published 2007
Reprinted 2008, 2011

FUNDAMENTAL FORCES OF NATURE
The Story of Gauge Fields

ISBN-13 978-981-270-644-7
ISBN-10 981-270-644-5
ISBN-13 978-981-270-645-4 (pbk)
ISBN-10 981-270-645-3 (pbk)

Printed in Singapore.

Contents

Preface

In this book I want to tell the story of gauge fields, the messengers that transmit signals among elementary particles, enabling them to interact. They work in the quantum realm of quarks, the deepest level of the structure of matter we have reached so far.

The basic interaction at this level percolates upwards, through hierarchies of organizations, to the everyday world we live in.

On its way, the interaction appears in different guises — nuclear interaction, atomic interaction, and the classical electromagnetic interaction that rules our everyday world. But these are facets of the same basic interaction.

The idea of "gauge" first appeared in electromagnetism. At the level we speak of, however, it is inextricably tied with the "quantum phase", that abstract attribute that distinguishes the microscopic world from the macroscopic, and that, incidentally, empowers new technologies of the 21st century, such as atom lasers and quantum computing.

The story of gauge fields is the story of our quest for the fundamental law of the physical world. It is the story of theoretical physics, from the time when Newton defined the meaning of force through his law of motion. To tell the story, we have to start from that beginning, for the thread is continuous and unbroken.

This book is not about the history of gauge theory, however. Our main goal is to introduce the idea behind gauge theory. We cover people and events relevant to gauge theory; but the order of narration follows ideas, rather than history.

Theoretical physics has given us a *true* understanding of the physical world. To quantify its achievement, we only have to note that theory agrees with experiment to one part in a trillion, in the most up-to-date measurement of the electron's magnetic moment.

Our greatest wonderment is to be reserved for the fact that our theories are not only true, but also beautiful. Theoretical physics is truly blessed, in that the quests for truth and beauty coincide. At the end of the book, we draw on what we have learned to offer a possible explanation of this remarkable coincidence.

Kerson Huang
January 2007

Introduction

In the everyday world, the most immediate interaction we are aware of is gravity. It makes heavenly bodies go round. It keeps us from jumping into orbit. To walk upright is to defy it. Paradoxically, it is the least understood of all interactions.

Better understood is the electromagnetic interaction. It underlies atomic structure and chemical reactions, thus giving us light and fire. It is responsible for almost all the happenings in our daily life.

James Maxwell's 1860 classical theory of electromagnetism is a "gauge theory". That means the basic field can freely change its "gauge" without affecting physical quantities. This principle of "gauge invariance" dictates the form of the electromagnetic interaction.

In 1954, Chen-Ning Yang and Robert L. Mills created what is now known as Yang–Mills gauge theory, through a creative generalization of Maxwell's theory. For almost twenty years, however, it remained in hibernation as a beautiful but useless mathematical exercise. That changed in the 1970s when, after breath-taking discoveries in particle physics, both experimental and theoretical, it was called upon to unify the electromagnetic and weak interactions. It now serves as the foundation of the Standard Model of elementary particles.

All the non-gravitational interactions we know of — strong, electromagnetic, weak — are described by Yang–Mills gauge theories. Einstein's theory of gravitation is a gauge theory of a sort; but it falls outside of the Yang–Mills mold, because of a close-knitting between space-time and inner structure.

The theory of gravitation deals with phenomena on a cosmic scale, whereas Yang–Mills theory is concerned with the opposite end — the smallest scale conceivable. Someday the two will meet, when we come to grips with what is inside that perceived singularity we call the "black hole". But this lies in the great unknown beyond the scope of this book.

The language of physics is mathematics, and we cannot avoid it, even in a semi-popular exposition such as this book. That does not mean, however, that the reader has to understand the equations. One could get the flavor of what is being discussed without the equations, just as one could enjoy a foreign movie without the subtitles.

Some readers, on the other hand, may want to see more equations. They will find them in the following technical books by the author:

- *Quarks, Leptons, and Gauge Fields*, 2nd edn. (World Scientific, Singapore, 1992);
- *Quantum Field Theory: From Operators to Path Integrals* (Wiley, New York, 1998).

1

What Makes the World Tick?

1.1. Motion

We see motion all around us. Leaves fall; waves break; heavenly bodies move.

What causes motion?

The answer is interaction. Interaction makes the world tick.

If there were no interactions, bodies would stand still, or move with unchanging velocity. Any change requires force, and that means interaction. Newton's law, the foundation of classical mechanics, states

$$\mathbf{F} = m\mathbf{a}.$$

Here, $\mathbf{F}$ is the force acting on a body, m is the inertial mass of the body, and $\mathbf{a}$ is its acceleration — the rate of change of the velocity. We can use this equation in two ways:

- as definition of force;
- as equation of motion.

In the first instance, we obtain the force $\mathbf{F}(\mathbf{x})$ by measuring the acceleration of the body at position $\mathbf{x}$. The force can be represented by a table of data, or by a force law we deduce from the data.

When the force is given, Newton's equation takes the form of a differential equation that can be solved, either analytically using calculus, or through numerical integration on a computer:

$$\ddot{\mathbf{x}} = \frac{\mathbf{F}(\mathbf{x})}{m}.$$

Galileo Galilei
(1564–1642)

Fig. 1.1 Galileo dropped two balls from the top of the leaning tower of Pisa, one light, the other heavy. They hit the ground simultaneously, showing that the acceleration due to gravity is independent of mass.

An overhead dot denotes time derivative. Thus, $\dot{x}$ denotes velocity, and $\ddot{x}$ is acceleration. Time has entered the picture, and the equation describes dynamical evolution.

1.2. Gravitation

The earliest known interaction is gravity. As legend has it, Galileo dropped two balls from the top of the Leaning Tower of Pisa, one heavy, the other light. They hit the ground simultaneously, showing that the acceleration due to gravity is independent of the mass. That

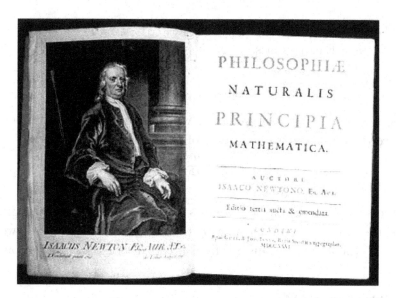

Fig. 1.2 Isaac Newton (1643–1727) laid the foundation of theoretical physics in his *Principia Mathematica* (1637).

is, $a = g$.[1] Newton's law then identifies mg as the force of gravity acting on a body. When this is substituted into the second form, the mass m cancels, and we get $\ddot{x} = g$. We can use this equation to calculate the path of a projectile, such as a golf ball.

The force due to gravity is approximately constant only near the surface of Earth. When you leave the surface, the force decreases inversely as the square of the distance from the center of Earth.

Newton's law of universal gravitation gives the force of attraction between any two bodies:

$$\text{Gravitational force} = \frac{\gamma m m'}{r^2},$$

where r is the distance between their centers, m and m' are their respective masses, and γ is the *gravitational constant*.[2]

---

[1] The constant g is called "acceleration due to gravity", or simply but misleadingly "g-force". Its value is 9.8 m s^{-2}, or 32 ft s^{-2}.

[2] The value of the gravitational constant is $\gamma = 6.670 \times 10^{-11}$ m^3 kg^{-1} s^{-2}.

The Earth's pull on a person can be obtained by putting

m = Earth's mass,

m' = Person's mass,

r = Distance between person and center of Earth.

Thus, r is very nearly the radius of Earth, even for a high jumper; whence the approximate constancy of the acceleration of gravity:

$$g = \frac{\gamma m}{R^2},$$

where R is Earth's radius.

The same inverse-square law gives the force between Jupiter and Mars, the force acting on a comet by the Sun, and indeed on any two masses in the universe. This is why it is called *universal* gravitation.

1.3. The force field

A mass m exerts a gravitational force on any other mass, proportional to the latter's mass. The force per unit mass is called the gravitational field:

$$\text{Gravitational field} = \frac{\gamma m}{r^2}.$$

Any other mass at a distance r from it will feel a force equal to this field times its mass.

In a sense the mass alters the property of space, for it creates a force field permeating all space. The *field* is to become a central concept in modern physics.

1.4. Equivalence principle

The mass m appears both as a measure of inertia, and a measure of field strength. These two roles are conceptually distinct, and we should really denote them with different symbols:

- The inertial mass m_{inertia} is the quantity appearing in

$$F = m_{\text{inertia}} \, a \,.$$

Albert Einstein (1879–1955)

Fig. 1.3 Some three hundred years after Gallileo's Pisa experiment, Einstein explained it in terms of the geometry of space-time, in his theory of general relativity.

It measures the body's response to an external force.

- The gravitational mass m_{grav} appears in $\gamma m_{\mathrm{grav}}/r^2$, and measures the field strength it produces.

Experimentally, they have the same numerical value:

$$m_{\mathrm{inertia}} = m_{\mathrm{grav}} .$$

This is known as the *equivalence principle*, and appears to be accidental.

Einstein could not accept the accidental explanation. He held that the two masses can be considered equivalent only when their defining concepts are shown to be equivalent. In 1917, nearly three hundred years after Galileo's experiment, he turned the accident into an imperative through the theory of general relativity.

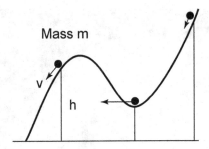

Fig. 1.4 In a roller coaster, kinetic energy $\frac{1}{2}mv^2$ and potential energy mgh convert into each other during the ride, but their sum remains constant.

In Einstein's general relativity, mass generates curvature in space-time. A body in its neighborhood simply rolls along a groove in curved space, following the shortest possible path (a geodesic). Thus, the mass has no bearing on motion in a gravitational field.

The actual curvature of space-time is very slight, and discernible only over cosmic distances. At relatively small scales, such as in the solar system, or even in galaxies, ordinary Netwonian mechanics is quite adequate.

1.5. Energy

A body has more "motion" when it goes faster, and a measure of the vigor is the kinetic energy

$$\text{Kinetic energy} = \frac{1}{2}mv^2,$$

where v is the velocity. When the body moves in a force field, the velocity changes from point to point.

For example, a roller coaster moves under gravity, at varying heights constrained by the track. The velocity is small near the top, and large near the bottom, as illustrated in Fig. 1.4.

We can define a potential energy mgh, where h is the height above ground. When added to the kinetic energy, we obtain a constant total energy, when friction is neglected:

$$\text{Total energy} = \text{Kinetic energy} + \text{Potential energy}.$$

This relation is known as the conservation of energy. The speedup and slowdown of a roller coaster signifies the conversion of potential energy to kinetic energy and *vice versa*.

1.6. Momentum

Momentum is defined as mass times velocity:

$$\text{Momentum} = m\mathbf{v}.$$

Newton's law says force is the rate of change of momentum. Thus, the momentum remains constant in the absence of force. This underlies the intuitive notion that momentum is what keep things on the move.

If a system is composed of more than one body, then each body has an individual momentum, and their sum is called the total momentum:

$$\text{Total momentum} = m_1\mathbf{v}_1 + m_2\mathbf{v}_2 + \cdots.$$

When there is no overall external force acting on the system, the sum of the internal forces must be zero, and the total momentum is conserved. If two particles collide in free space, their individual momenta will suffer changes, but the sum of the momenta must be the same before and after the collision.

1.7. Least action

The magic formula $F = ma$ explains the classical world.

Why is it true?

To properly pose the question, consider the motion depicted schematically in Fig. 1.5. The solid line represents a particle's actual path, which is governed by Newton's equation. The dotted lines represent other "virtual" paths with the same endpoints. How does

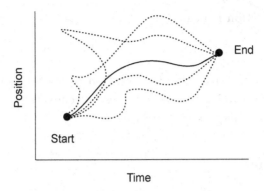

Fig. 1.5 A particle travels between two fixed endpoints. The solid curve is the correct path dictated by Newton's equation. It is singled out of all "virtual" paths (dotted curves) as one with the least "action".

the particle pick the correct path from the infinite number of virtual paths?

Joseph-Louis Lagrange answered this question with the *principle of least action*,[3] as follows. First, consider the quantity now known as the "Lagrangian":

$$\text{Lagrangian} = \text{Kinetic energy} - \text{Potential energy.}$$

We can calculate it along any virtual path. The "action" of the path is the Lagrangian accumulated over the entire path:

$$\text{Action of path} = \int_{\text{Path}} dt \ \text{Lagrangian.}$$

As we vary the path, the corresponding action changes. The correct path is that which minimizes the action.[4]

---

[3]The principle of least action had been proposed in various forms by Pierre Fermat (1601–1665), Pierre-Louis Moreau de Maupertuis (1698–1759), and Leonhard Euler (1707–1783).
[4]Actually, the sign of the action is immaterial, and the action could be maximal instead of minimal. For this reason purists prefer the name "principle of stationary action".

Joseph-Louis Lagrange   Pierre-Simon Laplace   William Rowan Hamilton
(1736–1813)   (1749–1827)   (1805–1865)

Fig. 1.6 Unlocking the power and beauty of Newtonian mechanics.

An early philosophical underpinning of the principle came from Laozi:[5]

Least action achieves all actions.

1.8. Newton canonized

Lagrange paved the way for William Hamilton, who based his approach on what we now call the "Hamiltonian":

Hamiltonian = Kinetic energy + Potential energy.

Its value is none other than the total energy, but the formalism requires that the Hamiltonian be expressed in terms of "canonical variables" — the coordinate q and its "canonically conjugate" momentum p. Accordingly we write it as $H(p,q)$. Newton's law is now recast in Hamilton's *canonical equations*:

$$\dot{q} = \frac{\partial}{\partial p} H(p,q),$$

$$\dot{p} = -\frac{\partial}{\partial q} H(p,q).$$

---

[5] 老子 道德经 (*Dao De Jing*, ca. 500 B.C.): "无为而无不为."

Urbain Le Verrier　　　　John Couch Adams　　　　Johann G. Galle
(1811–1877)　　　　　　(1819–1892)　　　　　　(1812–1910)

Fig. 1.7 Truimph of Newtonian mechanics: prediction and discovery of the planet Neptune.

The Lagrangian and Hamiltonian formulations are equivalent. The most succinct way to specify a system is to give its Lagrangian or Hamiltonian.

1.9. The mechanical universe

The correctness of Newtonian mechanics had been confirmed over and over in celestial mechanics, through the effort of Pierre Simon Laplace and others. The crowning moment was surely the prediction and discovery of a heretofore unknown planet — Neptune. Its existence was deduced independently by Urbain Le Verrier and John Couch Adams, from perturbations in the orbit of Uranus. A letter from Le Verrier containing the predicted planet's coordinates reached Johann Galle on September 23, 1846. The same evening, Galle wangled observation time on the Berlin telescope. Pointing it to the predicted position, he found Neptune.

The understanding of the the universe seemed complete. Laplace said that, given the positions and velocities of all the stars at any one instant, he will be able to calculate, in principle, the history of the universe for all times. The ability to quantitatively understand natural phenomena led to profound philosophical shifts.

The following exchange reportedly took place during a meeting of Laplace and Lagrange with Napoleon Bonaparte (1769–1821)[6]:

Napoleon: How is it that, although you say so much about the Universe, you say nothing about its Creator?
Laplace: No, Sire, I had no need of that hypothesis.
Lagrange: Ah, but it is such a good hypothesis: it explains so many things!
Laplace: Indeed, Sire, Monsieur Lagrange has, with his usual sagacity, put his finger on the precise difficulty with the hypothesis: it explains everything, but predicts nothing.

Laplace may think that he was able to predict everything; but his was a mechanical universe. An essential ingredient of the real universe was not yet considered: electromagnetism.

---

[6]A. De Morgan, *Budget of Paradoxes* (Longmans, Green, London, 1872).

2

Electromagnetism

2.1. Electric field

We know about electromagnetic interactions through electric and magnetic phenomena in everyday life. To understand these phenomena, we need to postulate a new attribute of matter called *electric charge*.

Coulomb's law states that two stationary electric charges exert a force on each other inversely proportional to the square of their separation. This is similar to the gravitational force between two masses, except for two things. First, the electric force is stronger by a fantastic order 10^{35}. Second, the electric charge can be either positive or negative, so that like charges repel each other, while opposite charges attract each other. Two opposites charges at the same position will neutralize each other.

Charles A. Couloumb (1736–1806) André Marie Ampère (1775–1836) Hans Christian Oersted (1777–1851) Jean-Baptiste Biot (1774–1862)

Fig. 2.1 Pioneers in electricity and magnetism.

Just as mass acts as a source of gravitational field, electric charge
is a source of electric field. A point charge q creates an electric field
pointing radially away from itself, with a magnitude inversely proportional to the squared distance from the charge. This is called the
Coulomb field:

$$\text{Coulomb field} = \frac{q}{r^2}.$$

Another charge q' in this field experiences a radial force equal to q'
times the field. The force can be repulsive or attractive, depending
on whether the sign of q' is the same or opposite to that of q.

2.2. Lines of force

We can picture the electric field by drawing "lines of force" tangent to
the field direction at each point of space, with a line density proportional to the field. More precisely, the electric field is the "line flux",
defined as the number of lines crossing a unit area perpendicular to
the direction of the field.

Electric lines of force "emanate" from positive charges, and are
"absorbed" by negative charges. They never break, and never cross
each other. A test charge placed in the electric field will move along
a line of force, like a fluid element moving along a streamline, with
acceleration proportional to the local line flux.

If we draw a sphere of radius r about an electric charge, the surface
area of the sphere will increase with r like r^2. Since the electric falls
off like r^{-2}, the number of lines piercing the surface of the sphere is
a constant that depends on the charge. This geometrical property,
known as *Gauss' law* is equivalent to Coulomb's inverse-square law.

2.3. Multipoles

Since the electric charge can be either positive or negative, we can
construct a hierarchy of elementary charge structures called multi-
poles:

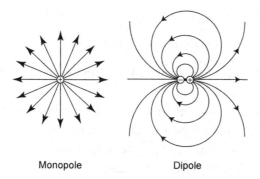

Monopole   Dipole

Fig. 2.2   Electric lines of force from monopole and dipole.

- A single charge q is an electric monopole. Its electric field decreases with distance like r^{-2},
- Two equal an opposite monopoles form a dipole. At large distances the field decreases like r^{-3},
- Two equal and opposite dipoles make a quadrupole. The field at large distances behaves like r^{-4},

and so forth. If you put a mixture of these things inside a "black box", you can reproduce any pattern of electric field outside, and no one will know what's really inside without opening the box.

Figure 2.2 shows the lines of force produced by a monopole and a dipole.

2.4. Scalar potential

The potential energy of two charges q and q' with separation r is given by

$$\text{Potential energy} = \frac{qq'}{r}.$$

If we keep q fixed and move q' around, the latter experiences different forces at different locations, and consequently has different potential energies. The potential energy per unit charge is called the *potential*.

The potential due to a point charge is called the *Coulomb potential*:

$$\text{Coulomb potential} = \frac{q}{r}.$$

We call this a "scalar potential" to distinguish it from the "vector potential" introduced later.

A collection of charges set up a scalar potential ϕ that is the sum of the individual Coulomb potentials. If we sprinkle a charge density ρ in the field of these charges, the interaction energy density is given by the potential energy per unit volume:

$$\text{Electric interaction energy density} = \rho\phi.$$

2.5. Electric current

Charge is conserved. It can disappear from a certain point only by moving elsewhere, and a moving charge generates an electric current. A distribution of charges can flow like a fluid, with

$$\text{Current density} = \text{Charge density} \times \text{Velocity},$$

or, in symbols,

$$\mathbf{j} = \rho\mathbf{v}.$$

The amount of current diverging out of a volume must equal the rate at which charge is being depleted from the volume:

$$\text{Divergence of current density} = \text{Rate of decrease of charge density},$$

or,

$$\nabla \cdot \mathbf{j} = -\frac{\partial \rho}{\partial t}.$$

This is called the *continuity equation*, an expression of charge conservation.

2.6. Magnetic field

Our earliest acquaintance with magnetism came from the tendency of bits of iron to adhere to a lode stone. The ancient Chinese characters

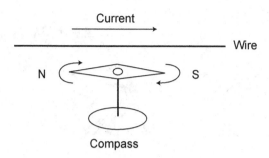

Fig. 2.3 In Oersted's pioneering experiment, an electric current causes a compass needle to deflect to one side. "How does the needle tell left from right?" asked Ernst Mach. See text for answer.

for magnet signify "maternal iron". We describe the phenomenon by picturing the existence of a magnetic field that exerts forces on particles of iron.

Hans Christian Oersted made the important discovery that an electric current generates a magnetic field, through the fact that it deflects a compass needle placed along side, as illustrated in Fig. 2.3.

Ernst Mach (1838–1916), a main opponent of the atomic theory of matter, found Oersted's experiment bewildering. How does the needle know which way to deflect, he wondered, when neither it nor the current-carrying can tell left from right?

But the compass does have a secret sense of left and right: it originates from spinning atoms making up the compass needle. The current also set up a magnetic field with particular handedness.

The *Biot–Savart law* says that the magnetic field lines of force form rings around the wire, with a direction given by the "right-hand rule": if you imagine grasping the wire with your right hand, with the thumb pointing along the direction of the current, then the field lines curl around the wire in the direction of your fingers.

There is no magnetic analog of a charge. The simplest source of a magnetic field is therefore not a "magnetic monopole" but a "magnetic dipole", which is equivalent to a current loop, as illustrated in Fig. 2.4. This makes magnetic phenomena seeming more complex than electric phenomena.

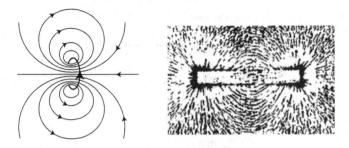

Fig. 2.4 Left: magnetic lines of force arising from a current loop, which represents a magnetic dipole. Right: lines of force from a permanent magnet made visible by iron filings. The permanent magnetic is a macroscopic dipole composed of microscopic atomic dipoles.

2.7. Vector potential

Since there are no magnetic charges, the magnetic field lines of force cannot terminate, and must run in closed rings. In mathematical terms,

- the magnetic $\mathbf{B}$ is "divergenceless": $\nabla \cdot \mathbf{B} = 0$.
- We can thus represent it as the "curl" of something: $\mathbf{B} = \nabla \times \mathbf{A}$.
- The quantity $\mathbf{A}$ is called the *vector potential*.

A magnetic field exerts a force on a current, and a current density $\mathbf{j}$ has interaction energy given by

$$\text{Magnetic interaction energy density} = -\mathbf{j} \cdot \mathbf{A} \,.$$

This is an important formula that we will use time and again in the future. It shows that charged particles are coupled to the magnetic field through the vector potential, which turns out to be the "gauge field" that is the central subject of this book.

2.8. Electromagnetic induction

According to Oersted's experiment, moving charges generate a magnetic field. In other words, a changing electric field begets a magnetic

Michael Faraday (1791–1867)

Fig. 2.5 Discoverer of electromagnetic induction and inventor of the dynamo.

field. Michael Faraday discovered the converse: a changing magnetic field generates an electric field. This is known as *electromagnetic induction*.

Specifically, a voltage difference develops across the ends of a metallic wire that is moving across a magnetic field. The magnitude of the induced voltage is proportional to the number of magnetic lines swept by the wire per second.[1] Faraday invented the dynamo based on this effect. His device is shown in Fig. 2.6, together with its modern descendant.

2.9. Maxwell's equations

Faraday synthesized the laws governing electromagnetic phenomena in four relations. They are stated in terms of the electric field **E** and

---

[1]From a microscopic point of view, the voltage difference results from free electrons in the metallic wire, being driven towards one end by an induced electric field.

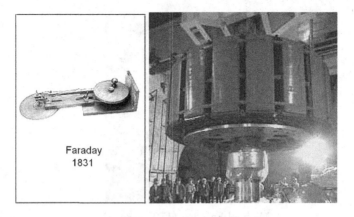

Fig. 2.6 Left: Faraday's dynamo (1831). Right: modern version at the hydroelectric power plant at the Three Gorges Dam, China (2006).

magnetic field **B**:

$$\text{Gauss' law:} \qquad \iint \mathbf{E} \cdot d\mathbf{S} = Q,$$

$$\text{No magnetic monopole:} \qquad \iint \mathbf{B} \cdot d\mathbf{S} = 0,$$

$$\text{Electromagnetic Induction:} \qquad \oint \mathbf{E} \cdot d\mathbf{x} = -\frac{1}{c}\frac{d\Phi}{dt},$$

$$\text{Ampere's law:} \qquad \oint \mathbf{B} \cdot d\mathbf{x} = \frac{1}{c}I.$$

The physical meaning of the equations are as follows:

- Electric flux out of any closed surface = Charge inside. (Equivalent to Coulomb's law.)
- Magnetic charge does not exist.
- Energy gained by test charge traversing any closed circuit $\propto$ rate of change of magnetic flux through circuit.
- Current generates magnetic field running in rings around the current.

The constant c is a unit that will turn out to be the velocity of light.

James Clerk Maxwell (1831–1879)

Fig. 2.7   Maxwell's equations define electromagnetic theory.

Maxwell noticed that the last law is not consistent with the conservation of charge, when the fields vary in time. He amended it, and expressed all the laws in differential form, by shrinking the closed surfaces and circuits to infinitesimal size.

The result is the set of Maxwell's equations that constitute the foundation of electromagnetism:

$$\nabla \cdot \mathbf{E} = 4\pi\rho \,,$$

$$\nabla \cdot \mathbf{B} = 0 \,,$$

$$\nabla \times \mathbf{E} = -\frac{1}{c}\frac{\partial}{\partial t}\mathbf{B} \,,$$

$$\nabla \times \mathbf{B} = \frac{4\pi}{c}\mathbf{j} + \frac{1}{c}\frac{\partial}{\partial t}\mathbf{E} \,.$$

Maxwell's addendum is the term $\frac{1}{c}\frac{\partial}{\partial t}\mathbf{E}$ in the last equation, known as the "displacement current". The presence of this term makes a momentous difference, for now there is the possibility for wave motion. These equations imply that a disturbance in the electromagnetic field will propagate at velocity c.

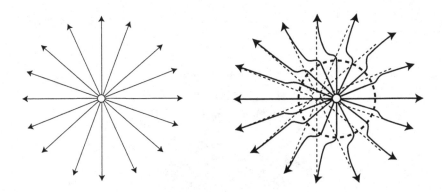

Fig. 2.8 Left panel: electric field lines of charge at rest. Right panel: charge is suddenly displaced a small distance and then stops. The field lines of force near the charge will move with it, but those far away will do so only after a time delay. The signal of change propagates as a spherical wave front. This represents a pulse of radiation.

2.10. Radiation

If we displace an electric charge suddenly, the electric field close to the charge will move with it. The field far away, however, does not immediately know that the source has moved. The information propagates with speed c, and will arrive at different distances at different times. As depicted in Fig. 2.8, the old field pattern switches over to the new pattern in a narrow shell, which propagates outward at constant speed c. Very far from the charge, the disturbance becomes a spherical wave front transverse to the direction of propagation. The wave front has lost all memories of the source, and travels freely as a pulse of radiation.

Almost thirty years after Maxwell predicted electromagnetic radiation, Hertz created it in the laboratory, and found that c is numerically equal to the speed of light:

$$c = 2.998 \times 10^{10} \text{ cm s}^{-1}.$$

Heinrich R. Hertz (1857–1894)

Fig. 2.9   Hertz discovered electromagnetic waves and showed they propagate with the velocity of light.

The known electromagnetic spectrum is shown below:

Gamma ray	X-ray	Ultra-violet	Visible light	Infrared	Microwave	FM	TV	SW	AM
10^{-14}	10^{-12} 10^{-10}	10^{-8}	10^{-6} 10^{-4}	10^{-2}	1	10^2		10^4	

Wavelength (m)

A central question remains:

With respect to what should the velocity of light be measured? That is, what is the medium of electromagnetic waves?

3

The Vacuum is the Medium

3.1. The ether

What is the medium in which electromagnetic waves propagate?

Our experience with wave motion comes from things like water waves, which represent the motion of a material substrate. The wave concept is just a convenient characterization of the motion of the substrate.

It is therefore natural to assume that electromagnetic waves represent motions of a certain medium, and the velocity of light is measured with respect to it. The medium was named the "ether".

If the ether exists, the Earth must be moving through it, for it would be absurd to suppose that the Earth drags the entire ether with it when it revolves around the Sun. We can measure the velocity of an "ether wind" by measuring the difference in the velocity of light emitted along different directions on Earth. In a series of experiments designed to do this, Albert A. Michelson (1852–1931), assisted by Edward Morley (1838–1923), found a null result:

The velocity of light does not depend on the direction of emission.

To reconcile the Michelson–Morley experiment with traditional thinking, people went through contortions, saying that nature "conspires" to hide the ether from us, that our meter sticks shrinks as we move, etc.

Albert A. Michelson (1852–1931)

Fig. 3.1 "Ether wind" looked for and not found.

Einstein made the obvious but daring inference:

Light propagates with a speed constant to all observers; there's no medium but the vacuum.

This is a bold position to take, for it necessitates a sweeping change in our concept of space and time.

The velocity of an object depends on how fast you are moving with respect to it. If a train is traveling at 60 mph, and you are running along side at 10 mph, then it appears to you the train is moving at 50 mph. As you vary your speed, the train's apparent velocity will change proportionately.

If the speed of light is to be the same no matter how fast you run, some long-held beliefs must be revised.

3.2. Reference frames

To measure position at a certain instant of time, an observer needs:

- a coordinate frame (the x, y, z axes) to register his data;
- a clock to read the time t.

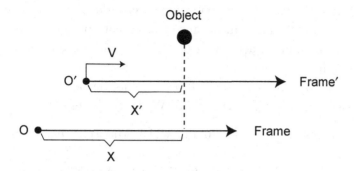

Fig. 3.2 An object has coordinate x in frame O, where it is at rest. In a moving frame O′, the coordinate becomes x'. The relation between x' and x is called a transformation law.

The reference frame is chosen as a matter of convenience, and generally varies from one observer to another. To relate data from different frames, we need a rule that translates the reading in one frame to another. This is called a *transformation law*.

A physical law must be independent of the observer. This means that it should be expressed by an equation that has the same appearance in all frames. We say that a physical law must be *covariant* with respect to the transformation law.

Consider two observers moving at a relative speed v, with coordinate frames as shown in Fig. 3.2. Common sense tells us that, the time t is the same in both frames, and that the position of an object measured by these observer, denoted x and x' respectively, differ by an amount determined by the relative velocity:

$$x' = x - vt \,,$$

$$t' = t \,.$$

This transformation law is called the *Galilean transformation*.

Newton's equation is covariant with respect to the Galilean transformation, because it can be expressed in vector form as $\mathbf{f} = m\mathbf{a}$. The components of the vectors $\mathbf{f}$ and $\mathbf{a}$ change from frame to frame, but the relation is the same in all frames.

Maxwell's equations, however, are not covariant with respect to the Galilean transformation, because the velocity of light c must be the same in all frames, according to Einstein's interpretation of the Michelson–Morley experiment.

There can be only one law of transformation, and that means Newton's equation should be revised. We must therefore:

- find the transformation law under which Maxwell's equations are covariant;
- amend Newton's equation so that it becomes covariant under the new transformation.

3.3. The light cone

We extend 3D space by adding time as a fourth dimension, and call a point in 4D space-time an "event". The space-time path traced out by a moving object is called a "world line". How should we define the "event distance" between two space-time points?

Let us choose an arbitrary origin, and denote the coordinates of a space-time event by the time t and the vector position $\mathbf{r}$. A ray of light ray emitted from the origin has a world line described by $\mathbf{r}^2 = (ct)^2$. This suggests that we define the event distance squared as

$$s^2 = (ct)^2 - \mathbf{r}^2,$$

so that a light ray is uniquely defined by the null world line corresponding to $s = 0$.

The collection of null world lines defines the *light cone*, which separates space-time into an "outside" and "inside", as depicted in Fig. 3.3. A body moving at less than light speed has a world line contained inside the light cone, while one traveling faster than light has a world line lying outside the light cone.

3.4. Lorentz transformation

How can we design a coordinate transformation that keeps the speed of light invariant? As a guide, we recall that a rotation is a linear

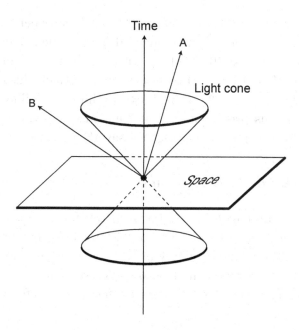

Fig. 3.3 The world line of a ray of light lies on the light cone. The world line A, which lies inside the light cone, corresponds to a body moving at less than light speed. World line B lies outside the light cone, and corresponds to motion faster than light.

coordinate transformation of the 3D spatial coordinates that keeps spatial distance between any two points invariant. We now seek a linear transformation in 4D space-time that preserves the event distance. Such a transformation will preserve the light cone, and hence the light speed. This was how Einstein posed the problem; the rest is algebra.

The result is the *Lorentz transformation*:

$$x' = \frac{x - vt}{\sqrt{1 - v^2/c^2}},$$

$$t' = \frac{t - vx/c^2}{\sqrt{1 - v^2/c^2}}.$$

For $v/c \to 0$, the transformation reduces to the Galilean transformation. For small v/c, deviations are proportional to $(v/c)^2$.

Considering that $c = 3 \times 10^{10}$ cm s^{-1} (186,000 miles per second), a supersonic jet plane reaches only one hundredth of one percent of the velocity of light: $v/c \approx 10^{-4}$. Thus, under ordinary circumstances, the fractional difference between the Lorentz and the Galilean transformation is less than 10^{-8}, or one part in a hundred million.

The momentous prediction is that Newton's law will fail, when $(v/c)^2$ grows to an appreciable fraction, say 1% or more.

3.5. Relativity of space and time

Hendrik Antoon Lorentz (1853–1928) wrote down the transformation law for Maxwell's equations, which would later bear his name. This brilliant formula was unfortunately muddled because he thought the time appearing there was some kind of "effective time".

Einstein realized that the transformation called for a fundamental recognition:

Motion mixes space and time.

He called his reformulation of space-time concepts the *theory of special relativity*, because the distinction between space and time is not absolute, but "relative". The theory is "special" because it only deals with frames moving at constant velocities.

If I am standing still, and you sail by at velocity v, your clock will not keep the same time as mine. The time you read in your rest frame is called your "proper time". This is a Lorentz-invariant quantity, because the instruction for finding it is the same for all observers: *move with that clock and read it.*

The Lorentz transformation implies that your proper time $d\tau$ is related to my proper time dt through the relation

$$d\tau = dt\sqrt{1 - v^2/c^2}\,.$$

Since $d\tau$ is smaller than dt, your motion causes your clock to run slower than mine, from my point of view. At "warp speed", your clock will stall completely, and you will never grow old, according to me.

The situation is symmetrical: from your point of view, I am the one who is moving, and my clock appears to run slow.

3.6. Four vectors

Covariance can be conveniently stated in terms of vectors. Equations stated in terms of ordinary vectors (3-vectors) are covariant under rotation. Similarly, an equation is automatically covariant under the Lorentz transformation, if it can be expressed in terms of 4-vectors.

A 3-vector has 3 components that transform under rotation like the coordinates x, y, z. A 4-vector has 4 components that transform under a Lorentz transformation like the space-time coordinates ct, x, y, z:

$$\text{3-vector:} \quad x^i = \{x, y, z\} \quad (i = 1, 2, 3)$$

$$\text{4-vector:} \quad x^\mu = \{ct, x, y, z\} \quad (\mu = 0, 1, 2, 3).$$

For simplicity, we denote the 4-vector as x instead of x^μ, when no confusion arises.

The geometry of 4D space-time is very different from that of 3D space, in that the squared distance $(ct)^2 - \mathbf{r}^2$ is not necessarily positive. This makes it necessary to distinguish two forms of 4-vectors, the "contravariant" and the "covariant". By definition, x^μ is contravariant. Its covariant form x_μ, written with a subscript instead of superscript, is obtained by reversing the signs of the spatial components:

$$x_\mu = \{ct, -x, -y, -z\} \quad (\mu = 0, 1, 2, 3).$$

The invariant product of two 4-vectors A and B is $A \cdot B = A^\mu B_\mu$, where the repeated index μ is automatically summed over $0, 1, 2, 3$ — a short hand initiated by Einstein, called the "summation convention". The space-time separation squared is the invariant product of x with itself: $s^2 = x \cdot x$.

3.7. $E = mc^2$

To recast Newton's law in covariant form, we first extend the momentum to a 4-vector:

$$p^\mu = m_0 \frac{dx^\mu}{d\tau},$$

where m_0 is the mass at rest, called the *rest mass*. When the velocity is small, τ reduces to t, and the spatial components p^i reduces to the familiar momentum. Putting $d\tau = dt\sqrt{1 - v^2/c^2}$, we can rewrite the 4-momentum as

$$p^\mu = m \frac{dx^\mu}{dt},$$

where the effective mass is

$$m = \frac{m_0}{\sqrt{1 - v^2/c^2}}.$$

This depends on the velocity, and approaches infinity as the velocity approaches that of light. Thus, we can never reach light speed, because the effective inertia keeps growing.

Newton's original law is now replaced by the covariant generalization

$$\frac{dp^\mu}{d\tau} = f^\mu,$$

where f^μ defines the 4-force.

The energy is the time component of the 4-momentum: $E = cp^0$. Thus,

$$E = \frac{m_0 c^2}{\sqrt{1 - v^2/c^2}}.$$

For small velocities, this reduces to $E \approx m_0 c^2 + \frac{1}{2} m_0 v^2$, which is the old kinetic energy, except for the constant term $m_0 c^2$. This says we assign a rest energy $m_0 c^2$ to a body. (We could subtract this from the definition, but it is more convenient to keep it.) In terms of the effective mass, then, we have

$$E = mc^2,$$

which is perhaps the single most famous equation in physics.

The Lorentz transformation becomes singular as the velocity approaches that of light. We are forever trapped inside the light cone. We can approach light speed, but never attain it.

3.8. Faster than light?

Is there a world outside the light cone, where everything moves faster than light? We will never know, according to special relativity, because bodies in that world can slow down and approach the light cone, but never reach it. In order to communicate with that world, we would need Lady Bright of the famous limerick:

> There was a lady named Bright,
> Who could travel faster than light.
> She went out one day,
> In a relative way,
> And came back the previous night.

3.9. Maxwell's true form

The key to the covariance of Maxwell's equations is how the vector potential transforms under a Lorentz transformation. The vector potential $\mathbf{A}$ and scalar potential ϕ together form a 4-vector

$$A = \{\phi, \mathbf{A}\}.$$

This is because its source, the current and charge densities, form a 4-vector current density

$$j = \{c\rho, \mathbf{j}\}.$$

The term "vector potential" will now mean 4-vector potential.

The electric and magnetic fields are components of a field tensor derived from the vector potential

$$F^{\mu\nu} = \partial^\mu A^\nu - \partial^\nu A^\mu.$$

We can display all the components in a table:

$$F^{\mu\nu} = \begin{pmatrix} 0 & -E^1 & -E^2 & -E^3 \\ E^1 & 0 & -B^3 & B^2 \\ E^2 & B^3 & 0 & -B^1 \\ E^3 & -B^2 & B^1 & 0 \end{pmatrix} ,$$

where rows and columns are numbered $0, 1, 2, 3$. Under a Lorentz transformation, the electric and magnetic fields mix, and behave in a manner much more complicated than that of the vector potential.

The field tensor has a dual $\tilde{F}^{\mu\nu}$, obtained through the replacement $\mathbf{E} \to \mathbf{B}$ and $\mathbf{B} \to -\mathbf{E}$. This can be expressed as

$$\tilde{F}^{\mu\nu} = \frac{1}{2} \epsilon^{\mu\nu\alpha\beta} F_{\alpha\beta} ,$$

where $\epsilon^{\mu\nu\alpha\beta}$ is the "completely antisymmetric tensor of rank 4".[1] In covariant form, Maxwell equations consist of two equations:

$$\partial_\mu F^{\mu\nu} = -\frac{4\pi}{c} j^\nu ,$$

$$\partial_\mu \tilde{F}^{\mu\nu} = 0,$$

3.10. The gauge field

From the point of view of covariance, the vector potential is the basic variable. As we shall see in the next chapter, it is also the fundamental quantity in the principle of least action.

However, the definition $F^{\mu\nu} = \partial^\mu A^\nu - \partial^\nu A^\mu$ does not uniquely determine A^μ. We can add to it any 4-vector of the form $\partial^\mu \chi$, where χ is any function of space-time.[2] The transformation,

$$A \to A + \partial \chi ,$$

is called a *gauge transformation*, and χ is called the gauge function. The fact that A is ambiguous up to a gauge transformation earns

---

[1]The tensor $\epsilon^{\mu\nu\alpha\beta}$ can have only 3 values: $0, \pm 1$. It is zero unless the indices are some permutation of $\{0123\}$. It is 1 if the permutation is even, and -1 if odd.
[2]The extra term does not contribute to $F^{\mu\nu}$ because $\partial^\mu(\partial^\nu \chi) - \partial^\nu(\partial^\mu \chi) = 0$.

it the name "gauge field". All physical quantities depend only on the electric and magnetic fields, and are therefore "gauge invariant", i.e. independent of the gauge function.

Introducing the gauge field immediately satisfies the second of Maxwell's equations: $\partial_\mu \tilde{F}^{\mu\nu} = 0$. The first reduces to

$$\Box A = \frac{4\pi}{c} j \,,$$

where $\Box = \partial \cdot \partial = \frac{1}{c^2} \frac{\partial^2}{\partial t^2} - \mathbf{\nabla}^2$ is the Lorentz-invariant wave operator. This says that the current density j is the source of the gauge field, and the field can propagate as a traveling wave with constant velocity c.

The interaction energy density between matter and the electromagnetic field is the sum of electric and magnetic contributions $\rho\phi - \mathbf{j} \cdot \mathbf{A}$, which can be neatly expressed in the Lorentz-invariant form

$$\text{Interaction energy density} = j \cdot A \,.$$

As we shall see in the next chapter, gauge invariance dictates the form of this interaction.

3.11. Who wrote these signs

Equations of physics have a beauty of their own as graphics. They also confer power on all who understand them, and this fact enhances their impact. Ludwig Boltzmann (1844–1906) paid tribute to Maxwell's equations by quoting from Geothe's *Faust*:

> Was it a god who wrote these signs?
> That have calmed yearnings of my soul,
> And opened to me a secret of Nature.

Maxwell's equations have gone through different representations, each stressing a particular aspect. Figure 3.4 displays the various forms imprinted on college T-shirts corresponding to levels of sophistication, from sophomore, senior, to graduate student. The Faraday form conveys a global picture of lines of force. Maxwell's differential

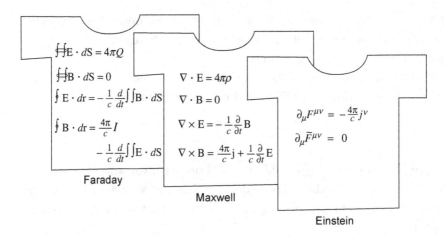

Fig. 3.4   College T-shirts with Maxwell's equations, as sported by sophomore, senior, graduate student.

form stresses the local effects of electric and magnetic fields. Finally, Einstein's covariant form brings out the true essence.

3.12. Lorentz and Einstein

The crux of the Lorentz transformation is that space and time get mixed up when you move — a wee bit only, if your velocity is much less that of light; but mix they must. Ironically, this point was lost on its originator Hendrik Lorentz, who confessed in hindsight:[3]

> The chief cause of my failure was my clinging to the idea that only the variable t can be considered as the true time, and that the local time t' must be considered no more than an auxiliary mathematical quantity.

The mathematician Henri Poincaré (1854–1912) wrote about the principle of covariance, which he called the "principle of relativity"; but it had no physical relevance, because he did not understand the "relativity" of simultaneity.

---

[3]A. Pais, *Sublte is the Lord, Biography of Einstein* (Oxford University Press, 2005), p. 167.

Fig. 3.5 Albert Einstein (1879–1955) and Hendrik Antoon Lorentz (1853–1928) in Leiden, 1921. (Source: Museum Boerhaave Leiden.)

P. A. M. Dirac (1902–1984) had this to say about the creation of special relativity:[4]

Any of you who have studied relativity must surely have wondered why it was that Lorentz succeeded in getting correctly all the basic equations needed to establish the relativity of space

[4]P. A. M. Dirac, *The Development of Quantum Theory*, J. Robert Oppenheimer Memorial Prize acceptance speech, Center for Theoretical Studes, University of Miami, 1971 (Gordon and Breach, New York, 1971), p. 13.

and time, but he just was not able to make the final step establishing relativity. He did all the hard work — all the really necessary mathematics — but he was not able to go beyond that, and you will ask yourself, "Why"?

I think he must have been held back by fears, some kind of inhibition. He was really afraid to venture into entirely new ground, to question ideas which had been accepted from time immemorial.

It needed several years and the boldness of Einstein to take the necessary step forward and say that time and space are connected. What seems to us nowadays a very small step forward was very difficult for the people in those days.

4

Let There be Light

4.1. Local gauge invariance

Accelerated charges emit light through its electromagnetic coupling, which is described by the interaction energy density $j \cdot A$.

Why?

Is there a deeper principle at work that determines the form of the interaction?

The answer is yes, and the principle is *local gauge invariance*.

The fundamental object in electromagnetism is the gauge field A; but it is not directly observable, since it is defined only up to a gauge transformation, and thus not unique.

We can picture the gauge field as a tower of values, related to each other by gauge transformation. Such a tower is called a *fiber* in mathematics, and a fiber is attached to each space-time point, as depicted in Fig. 4.1. The collection of all fibers on space-time is called a *fiber bundle*. Under a local gauge transformation, the field slides along its fiber, independently at each space-time point. The physical world, however, must not be aware of the acrobatics:

> The Hamiltonian of the world must be invariant under local gauge transformations.

This is the *principle of local gauge invariance*, which, as we shall see, dictates the form $j \cdot A$.

It seems strange that nature should hold sacred something we cannot directly observe — the gauge freedom. Can this principle of

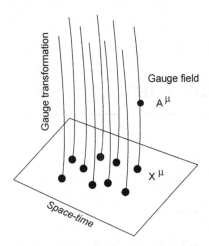

Fig. 4.1 The gauge field is represented by a fiber bundle over space-time. In a local gauge transformation, the gauge field slides along its fiber independently at each point of space-time. Local gauge invariance is the requirement that the physical world be blind to such acrobatics. This principle determines the form of the electromagnetic interaction.

gauge invariance be the last word? We do not know. If there is a deeper truth, physics has not yet discovered it.

4.2. A creation

With perfect vision of hindsight, let us derive the action of the world from "pure thought".

The action for a free relativistic particle is simplicity itself — the proper time spent in going from point a to point b:

$$S_{\text{particle}} = \text{Const.} \int_a^b d\tau \,,$$

where "Const." stands for some constant.[1]

For the electromagnetic field, which fills all space, the action is the space-time integral of a density, which must be Lorentz invariant

---

[1] For simplicity, we measure velocities in units of the velocity of light. Thus $c = 1$.

and gauge invariant. To look for the simplest combination of gauge fields that is both Lorentz invariant and gauge invariant, we reason as follows:

- The most obvious Lorentz-invariant combination is $A^\mu A_\mu$, but this is not gauge invariant.
- The field tensor $F^{\mu\nu} = \partial^\mu A^\nu - \partial^\nu A^\mu$ is gauge invariant. The simplest Lorentz invariant constructed from it is $F^2 = F^{\mu\nu}F_{\mu\nu}$, and this fulfills all our requirements.

Accordingly we take the action of the free electromagnetic field as

$$S_{\text{em}} = \text{Const.} \int F^2 \,,$$

where the integration extends over all space-time.

Now the interaction. It should be Lorentz invariant and gauge invariant. As the simplest possibility, we take it to be linear in the field and the particle coordinate. Under these conditions, the only thing we can write down is

$$S_{\text{int}} = \text{Const.} \int_a^b dx \cdot A \,,$$

where A is evaluated at the particle's position x. This does not look gauge invariant, but actually is, if the gauge function is the same at the endpoints.

We can rewrite:

$$S_{\text{int}} = \text{Const.} \int_a^b d\tau \frac{dx}{d\tau} \cdot A \,.$$

This says that the interaction energy is proportional to $\frac{dx}{d\tau} \cdot A$. Now, $\frac{dx}{d\tau}$ is the velocity of the particle, which is proportional to its current density j. Thus, we have the desired result

$$\text{Interaction energy} = j \cdot A \,.$$

This shows how gauge invariance determines the form of the interaction.

The complete world action is

$$S = -\frac{1}{4} \int F^2 + \int_a^b \left(-m d\tau + qA \cdot dx \right),$$

where we have determined the constants in terms of conventional definitions of rest mass m and electric charge q. This action will yield the complete relativistic equations of motion for particle and field, via the principle of least action. It represents the blueprint of the classical world, excluding gravitation.

We see that the structure of the world rests upon a few principles, as long as we know the correct choice of variables in a correct framework. History tells us, however, that the latter can come to light only through an arduous process of interaction and feedback between theory and experiment.

4.3. The gauge principle

We can now give a simple recipe to "turn on" the electromagnetic interaction.

From the world action, it is straightforward to calculate the Hamiltonian. The second term, in particular, yields the interaction Hamiltonian, which is the particle's energy in the electromagnetic field:

$$E = q\phi + \sqrt{(\mathbf{p} - q\mathbf{A})^2 + m^2}.$$

Rewriting the energy as p^0, the time component of the 4-momentum, we have

$$p^0 - q\phi = \sqrt{(\mathbf{p} - q\mathbf{A})^2 + m^2}.$$

Comparing this to the energy of the free particle

$$p^0 = \sqrt{\mathbf{p}^2 + m^2},$$

we see that the electromagnetic coupling appears through the 4-vector replacement

$$p \to p - qA.$$

Hermann Weyl (1885–1955)

Fig. 4.2 Weyl's gauge transformation, which failed to work in an old theory, found reincarnation in quantum mechanics.

This is known as the *gauge principle*. It "throws a switch" to turn the light on.

4.4. Hermann Weyl

The gauge transformation was introduced by Hermann Weyl in an attempt to reduce electromagnetism to world geometry, as Einstein had done for gravitation.

In Einstein's theory of general relativity, gravity is due to the curvature of space-time. In the presence of curvature, the direction of a vector becomes "non-integrable". That is, when the vector is transported parallel to itself along a close circuit, its angle is changed, by an amount proportional to the flux of the gravitation field linking the circuit.

Weyl theorized that the presence of an electromagnetic field makes a vector's length non-integrable. That is, the vector becomes "stretched" when it is parallel-transported around a close circuit that links electromagnetic field lines. He proposed a stretch factor

$$\exp\left(\frac{q}{\gamma}\oint dx \cdot A\right),$$

and called this a "gauge transformation" of the length scale. Here, A is the 4-vector potential of the electromagnetic field, q is the charge, and γ is a constant.

Einstein immediately pointed out that Weyl's idea is physically untenable, for, if our meter stick stretches every time we dance round a circle, then length has no meaning. Unlike direction, the length of a physical object must have a unique value. Disappointed but undaunted, Weyl declared from the luxurious lap of mathematics,

> When there is a conflict between beauty and truth, I choose beauty.

As it turns out, Weyl's idea was almost correct, but in an entirely different setting. His stretch factor turns out be a gauge transformation in quantum mechanics, with two important changes (Chap. 7):

- The constant γ is not real, but pure imaginary: $\gamma = i\hbar$, where $\hbar$ is Planck's constant divided by 2π. There is no stretching — the modulus of the factor is unity.
- The factor multiplies not the standard of length, but the quantum–mechanical wave function. Its business is not mensuration, but "entanglement".

The name "gauge transformation" stuck, but now "gauge" refers not to length scale but the quantum phase.

4.5. And there was light

The gauge principle shows us how to turn the light on. With this in mind, we write footnotes to The Book of Genesis:

In the beginning God created the heaven and the earth.

God designed the matter Hamiltonian $H(p, x)$.

And the earth was without form, and void; and darkness was upon the face of the deep. And the Spirit of God moved upon the face of the waters.

Something seemed missing. God pondered.

And God said, Let there be light:

Let $p \to p - qA$.

And there was light.

5

Heroic Age: The Struggle for Quantum Theory

5.1. Alien signals

We have enormous radio telescopes trained toward the sky, listening. We wait for that "intelligent" signal that may never come, from "aliens" that may not exist. But we did receive intelligent signals once from a unknown source. They were unsolicited, unwelcome, and deeply disturbing.

The signals came from light spectra emitted by atoms, at a time when we had mastered Newtonian mechanics, Maxwell's equations, and thermodynamics. These theories had explained all known phenomena. In the view of William Thomson (Lord Kelvin) of entropy fame,

> There is nothing new to be discovered in physics now. All that remains is more and more precise measurement.[1]

In that magnificent edifice that was classical physics, the atom appeared merely as a convenient metaphor. There was no hard evidence for its existence, and there were prestigious voices against it, notably from Ernst Mach (1838–1916) and Wilhelm Ostwald (1853–1932). True, the tide began to turn after 1905, due to the work of Albert Einstein and Marian Smoluchowski (1872–1917) on Brownian motion. Based on their suggestions, Jean-Baptiste Perrin (1870–1942) measured Avogadro's number in 1908. People began to admit that

---

[1]Address at the British Association for the advancement of Science (1900).

William Thomson (Lord Kelvin) (1824–1907)

Fig. 5.1   "There is nothing new to be discovered in physics now" (1900).

maybe matter has a "graininess"; but that should mean only a minor adjustment of our world view. When we finally acquired the ability to really "listen" to the atoms however, we were not prepared for what we heard.

Newton had decomposed sunlight into a spectrum of colors by passing it through a prism. More than a century later, Joseph Von Fraunhofer (1787–1826) passed it through a narrow slit, and found dark absorption lines in the spectrum of sunlight. Modern spectroscopy began in 1882, when Henry Rowland (1848–1901), first physics professor at Johns Hopkins University, invented a way to make good gratings. Within a few years, he was able to obtain a solar spectrum 50 feet in length. Soon it became routine to obtain good atomic spectra, which consist of series of lines corresponding to light emitted at various discrete frequencies.

The mathematician Johann Balmer (1825–1898) cracked the code of a hydrogen spectrum, now known as the Balmer series:

$$\text{frequency} = b\left(1 - \frac{4}{n^2}\right) \qquad (n = 3, 4, 5, \dots),$$

Henry A. Rowland	Johann J. Balmer
(1848–1901)	(1825–1898)

Fig. 5.2 Left: Rowland's grating produced a solar spectrum 50 feet long. Right: Balmer cracked the code of the hydrogen spectrum.

where b is a constant. If that's not an intelligent signal, I don't know what is.

Twenty years passed before we had a glimmer of what this formula meant. We had to wait for a picture of the atom to emerge from experiments. With J. J. Thomson's discovery of the electron in 1897, and Ernest Rutherford's discovery of the atomic nucleus in 1911, it became clear that an atom consists of electrons surrounding a small, heavy, positively charged nucleus.

5.2. Bohr's atom

In a flash of insight, Niels Bohr derived the Balmer formula in 1913, in a simplistic model that nevertheless captured the essence of the atom. He assumed that the electron in a hydrogen atom forms a standing wave about the central nucleus. Thus, the length of its orbit must be a multiple of the wavelength. This quantizes the orbits and their energies.

When an electron jumps from a higher orbit to a lower one, the energy difference E is released in the form of light, whose frequency

Fig. 5.3 J. J. Thomson (1856–1940) and Ernest Rutherford (1871–1937) elucidated the structure of the atom as a central nucleus surrounded by electrons. Photograph by D. Schoenberg, courtesy of AIP Emilio Segre Visual Archive (Bainbridge Collection).

ν is given through a formula of Planck and Einstein:

$$E = h\nu\,,$$

where h is Planck's constant:

$$h \approx 6.63 \times 10^{-27} \text{ erg-sec.}$$

The quantum jumps that give the Balmer series are indicated in the energy levels diagram in Fig. 5.4. The Bohr model explains the data, but raised many questions. In classical physics, an electron running around the nucleus will lose energy to radiation and

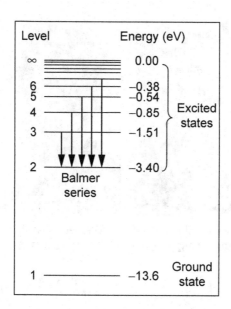

Fig. 5.4 Energy levels of the hydrogen atom in the Bohr model, given in electron volts (eV). Arrows indicate the quantum jumps that give rise to the Balmer series.

spiral into the nucleus in less than a microsecond. So an immediate question is,

What makes the electron's orbit stable?

Bohr:[2]

We are now in a new field of physics, in which we know that the old concepts probably don't work, because otherwise atoms wouldn't be stable. On the other hand, when we want to speak about atoms, we must use words, and these words can only be taken from old concepts, from the old language. Therefore we are in a hopeless dilemma.

[2]Recount by W. Heisenberg at the Conference on Contemporary Physics, Trieste, 1968, published in *From a Life of Physics* (World Scientific, Singapore, 1989), p. 37.

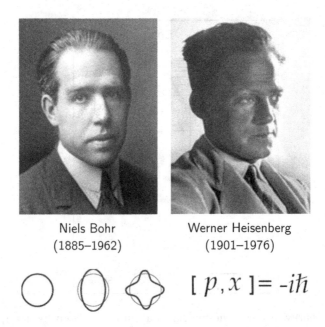

Niels Bohr
(1885–1962)

Werner Heisenberg
(1901–1976)

$$[\,p,x\,] = -i\hbar$$

Fig. 5.5 Trail blazers: Bohr with his orbits, and Heisenberg with his commutator.

Heisenberg:[3]

> The decisive step is always a rather discontinuous jump. You must really leave the old concepts and try something new, and then see whether you can swim, or stand, or whatever else; but in any case you can't keep the old concepts.

And Heisenberg made the jump.

5.3. Purely imaginary

Heisenberg's idea was to work only with observable quantities. This way, you avoid concepts like electron orbits. He studied a large amount of spectroscopic data, and came to the disturbing conclusion that the momentum and position of the electron are not commuta-

[3] *Op. cit.* p. 44.

tive. They should be represented by matrices, with the *commutation relation*

$$[p, x] = -i\hbar,$$

where the bracket symbol denotes the commutator: $[p, x] \equiv px - xp$, and $\hbar$ is Planck's constant divided by 2π:

$$\hbar = \frac{h}{2\pi} = 1.054 \times 10^{-27} \text{erg s.}$$

Most significantly, the commutator contains the purely imaginary number $i = \sqrt{-1}$.

For the first time in physics, theory ventures into a new dimension — the complex plane.

Heisenberg's commutation relation has since become the foundation of quantum mechanics; but Heisenberg felt unsure about it, and buried it inside his paper of 1925.[4] Dirac recalled:[5]

It was quite inconceivable that two physical things when multiplied in one order should not give the same result as when multiplied in the other order. It was thus most disturbing to Heisenberg. He was afraid this was a fundamental blemish in his theory and that probably the whole beautiful idea would have to be given up.

I received an early copy of Heisenberg's first work a little before publication, and I studied it for a while, and within a week or two I saw that the non-commutation was really the dominant characteristic of Heisenberg's new theory. It was really more important than Heisenberg's idea of building up the theory in terms of quantities closely connected with experimental results. So I was led to concentrate on the idea of non-commutation, and to see how the ordinary dynamics, which people had been using until then, should be modified to include it.

[4] W. Heisenberg, "Über quantentheoretische Umdeutung kinematischer und mechanischer Beziehungen", *Zeitschrift fúr Physik* **33**, 879–893 (1925).
[5] P. A. M. Dirac, *op. cit.*, p. 22.

At this stage, you see, I had an advantage over Heisenberg because I did not have his fears.

In 1926, Erwin Schrödinger proposed a wave equation, extending the idea of Louis De Broglie (1892–1987) on the wave nature of the electron. This *Schrödinger equation* has since become the working tool of quantum mechanics. Ironically, it was published grudgingly, as a watered down version of an original, more "beautiful" equation. According to Dirac:[6]

> De Broglie's ideas applied only to free electrons and Schrödinger was faced with the problem of modifying De Broglie's equation to make it apply to an electron moving in a field, in particular, to make it apply to electrons in atoms. After working on this for some time, Schrödinger was able to arrive at an equation, a very neat and beautiful equation, which seemed to be correct from a general point of view.
>
> Of course, it was necessary then to apply it, to see if it would work in practice. He applied it to the problem of the electron in the hydrogen atom and worked out the spectrum of hydrogen. The result that he got was not in agreement with experiment. That was most disappointing to Schrödinger. ... He abandoned the thing for some months, as he told me. And then, afterwards, when he had recovered from his depression somewhat, he returned to this work and noticed that if he applied his ideas with less accuracy, not taking into effects due to the relativistic motion of the electron, with this lesser accuracy, his theory agreed with observation.

5.4. Quantum mechanics

It was Dirac who formulated quantum mechanics as a consistent theory, and showed that the ideas of Heisenberg and Schrödinger

[6]P. A. M. Dirac, *op. cit*, p. 37.

were equivalent. The theory can be summarized as follows:

- The state of a system corresponds to a vector in an abstract "Hilbert space". The vectors ψ and $c\psi$ describe the same state, where c is a complex number.
- An observable like momentum is associated with an operator acting on state vectors. Measuring the observable in one of its eigenstates will yield the corresponding eigenvalue. Measuring the observable in a non-eigenstate will yield a statistical distribution of eigenvalues. To insure that the eigenvalues are real, the operators should be "hermitian".
- A classical theory can be "quantized" by converting the Hamiltonian $H(p, x)$ into an operator, using Heisenberg's commutation relation $[p, x] = -i\hbar$. This procedure is known as *canonical quantization*.
- The Hamiltonian is the generator of time evolution. This is expressed by the Schrödinger equation:

$$H\psi = i\hbar\frac{\partial\psi}{\partial t}\,.$$

Heisenberg's commutator makes momentum and position truly "canonical" variables, for they are the quantized ones. The astounding thing is that it goes off the real axis in the imaginary direction. Quantum mechanics is at once canonical and transcendental, in ways unfathomable in classical thought.

5.5. The wave function

We can represent operators in different ways that are equivalent. Heisenberg chose to represent them by matrices, while Schrödinger represented them by differential operators, identifying p as $-i\hbar\frac{\partial}{\partial x}$.

In Schrödinger's representation, the state representative ψ is a function of position and time called the wave function. It is a complex number, and cannot be observed directly, but it is what we calculate through the Schrödinger equation.

Max Born showed that the wave function is a "probability

Fig. 5.6 P. A. M. Dirac (1902–1984) formulated quantum mechanics as we know it, and wrote down a relativistic equation for the electron. Here, he was apparently giving a lecture on the quantum mechanics of the hydrogen molecule.

amplitude", in the sense that the squared modulus $|\psi|^2$ is the probability density of finding the particle at a certain position at a given time.

What makes ψ not directly observable is the complex phase. The relative phase between two wave function is observable, however, and the existence of the relative phase is what truly marks the difference between quantum mechanics and classical mechanics. (More in Chap. 6).

5.6. Quantum theory and relativity

Schrödinger's original "beautiful" equation that failed to work was based on relativistic covariance. The "watered-down" version that works, known to us as the Schrödinger equation, is a non-relativistic

Erwin Schrödinger   Max Born
(1887–1961)   (1882–1970)

Fig. 5.7   Schrödinger represents the state of a particle by a wave function, which Born interprets as probability ampliude.

approximation. The marriage between quantum theory and relativity raises deep issues not easily resolved.[7]

In 1928, Dirac wrote down a relativistic equation for an electron, with intrinsic spin 1/2 (in units of $\hbar$). To achieve consistency, he had to describe the electron with a 4-component wave function, with unexpected and far-reaching properties that opened the door to quantum field theory. (More in Chap. 11.)

5.7. Silly question

After all that, you will still ask

So, what makes the electron's orbit stable?

Answer: That's a silly question! There are no orbits, only quantum states that are solutions to the Schrödinger equation. An electron in the hydrogen atom is represented by a stationary cloud of charge distribution.

---

[7] Schrödinger's original equation is now recognized as describing a relativistic field theory of spin 0 particles, but it is called the Klein–Gordon equation.

The real question is how an electron emits light by jumping from one state to another.

To answer this question, we need to understand the electromagnetic interaction in quantum theory, and that requires some ground work. The answer will have to wait till Chap. 10.

6

Quantum Reality

6.1. The uncertainty relation

Quantum mechanics is such a radical departure from classical physics that the very notion of the state of a particle has to be changed. The position and momentum are now non-commuting operators, and you cannot have a simultaneous eigenstate of both. If the particle is completely localized, then the wave function is a sharp peak at a definite location, but the momentum is completely unknown. On the other hand, if the momentum is definite, then the wave function is a plane wave, and the position cannot be specified — the particle is everywhere.

If you strike a comprise by allowing the particle to have a range of positions in an interval Δx, and a range of momenta in an interval Δp, then the commutator says their product must be greater than $\hbar$:

$$\Delta x \Delta p > \hbar.$$

This is Heisenberg's *uncertainty relation*.

A wave packet with a spatial extension of ten thousandth of a millimeter, which would appear as a point-particle to us, can have a momentum defined to a tolerance of 10^{-22} cgs units, and we can safely say that it has a precise value. Planck's constant is so tiny on a macroscopic scale that, for all practical purposes, we can regard both position and momentum as well defined.

In the microscopic world there is uncertainty; but it refers only to the spread in the observed values of a physical quantity in noneigenstates. There is no uncertainty in the theory, either in mathematical formulation, or in its prediction of experimental outcomes.

If you lived in a world in which everything you could see was made up of regular polyhedra of at least a billion sides, you would characterize everything with a diameter, since they all look like spheres. One day, you were given a very powerful microscope, and looking through it, you discovered a small polyhedron with ten sides. You had never seen anything like this before, and "side" was not even in your dictionary.

The first question you raised was, naturally, "What's the diameter?"

"Well, it's uncertain."

6.2. Wave nature of matter

Schrödinger's wave function ψ is a complex number, and as such has both a magnitude and a phase:

$$\psi = Re^{i\theta}.$$

Max Born pointed out that R^2 gives the probability distribution of the particle's position. The phase factor leads to interference phenomena characteristic of waves.

The Schrödinger equation is a linear equation for ψ. This leads to the *superposition principle*, namely, the sum of two solutions is a solution. Suppose one solution describes state 1, and another solution describes a different state 2. When you add the two solutions, you get the wave function of a new state:

$$\psi = c_1\psi_1 + c_2\psi_2,$$

where the coefficients c_1, c_2 are complex numbers.

Probabilities are additive in a classical setting; but here the probability amplitudes add, not the probabilities. When you square the sum of amplitudes to obtain the probability, an extra "interference term" appears. Taking $c_1 = c_2 = 1$ for illustration, we have the probability distribution

$$R^2 = R_1^2 + R_2^2 + 2R_1 R_2 \cos(\theta_2 - \theta_1).$$

The first two terms represent the classical additive result, and the last term is the interference. Such interference phenomenon is common in water waves or electromagnetic waves. What is new here is that we are dealing with matter waves — waves of probability in a particle's position.

The quantum phase θ is a distinctive feature of quantum mechanics that has no analog in classical mechanics. When the quantum phase can be ignored, interference effects go away, and quantum mechanics reduces to classical mechanics.

Such a situation applies to a macroscopic body — such as a cat — whose energy levels are so closely spaced as to be a continuum in any practical sense. It is also constantly interacting with an environment having an enormous number of degrees of freedom. Consequently, its state is not a single eigenstate of energy, but a superposition of an enormous number of eigenstates, with relative phase angles fluctuating rapidly and randomly. Consequently, all interference effects average to zero.

6.3. Entanglement

A special kind of interference is *entanglement*, a term coined *by* Schrödinger.[1]

---

[1]Schrödinger gave the following somewhat opaque definition of entanglement, in "Discussion of Probability Relations Between Separated Systems," *Proc. Camb. Phil. Soc.* **31**, 555 (1935); **32**, 446 (1936):

When two systems, of which we know the states by their respective representatives, enter into temporary physical interaction due to known forces between them, and when after a time of mutual influence the systems separate again, then they can no longer be described in the same way as before, viz. by endowing each of them with a representative of its own. I would not call that one but rather the characteristic trait of quantum mechanics, the one that enforces its entire departure from classical lines of thought. By the interaction the two representatives [the quantum states] have become entangled.

Suppose two particles exist in an eigenstate of a certain observable, such as angular momentum, but neither particle is in an individual eigenstate of that observable. The two particles are said to be entangled.

The entangled state seems to have unusual and startling properties.

We can "force" one of the particles into an eigenstate of its own, by performing a measurement of the observable. Then the other particle must "collapse" into a corresponding eigenstate of its own, even though the particles may be far apart in space.

However, terms like "force into a state" and "collapse into a state" are just manners of speech. Entanglement refers to correlations in *simultaneous* measurements of the observable. Such correlations have been experimentally observed in small systems, and is the basis of quantum computing.

When extrapolated naively to the macroscopic domain, entanglement leads to nonsense.

Example: someone takes one look at your friend in Boston, and instantly you collapse in Hong Kong.

Fear not, for macroscopic bodies cannot exhibit quantum interference, as explained in the last section.

As a simple example of entanglement, consider two non-relativistic particles of spin 1/2, so that the spin state is either "up" or "down". We assume that the wave function of a particle can be factored into a spatial part and a spin part, and we deal only with the spin part.

There are two independent spin wave functions α and β, corresponding respectively to the up and down state. In the composite two-particle system, the total spin can have the values 1 or 0. For total spin 0, the wave function is proportional to

$$\alpha_1\beta_2 - \beta_1\alpha_2 \, ,$$

where the subscripts identify the particles. In this state, the total spin is definite, but the individual spins are not definite. All we know is that one spin is up, and the other is down, but we *cannot* know which is which. The two particles are "entangled".

Now, if we measure the spin of particle 1, the outcome has to be either up of down. That means the total wave function will be "forced" into becoming either the first term or the second term. Thus, if we find that particle 1 has up spin, then particle 2 must have down spin, and *vice versa*. Performing a measurement on one particle determines the state of the other particle, even though the two particles may be separated in space.[2]

There is no conceptual problem if we look upon the above as a description of the correlation between *simultaneous* measurements of the individual spins near each other. The results are not what you would expect classically, but that's quantum mechanics.

The confusion arises when we extend the reasoning to macroscopic objects, or to spin separated by large distances. It would then appear that there is some kind of "spooky action-at-a-distance", a phrase used by Einstein.

But the reasoning fails in both cases.

First, the reasoning fails for macroscopic objects, because they cannot exist in pure quantum states, as pointed out earlier.

Secondly, when the two spins are sufficiently far separated, the problem has to be treated relativistically, because the question of signal transmission becomes relevant. In relativistic quantum theory, however, one faces an immediate complication, namely spin becomes entangled with space-time, and the wave function is no longer factorizable.

From an experimental point of view, it is hard to entangle two spin far separated, because they are easily "dephased" by small perturbations. This problem remains a subject of research.

---

[2]Spatial information is not contained in the spin wave function. We have factored out the spatial part in a non-relativistic setting. Statements about the locations of the particles are only valid in a non-relativisitic setting.

6.4. All virtual realities

The quantum reality includes all virtual realities.

Every manifestation, no matter how fantastic, has its *chana*[3] of reality.

We can write an uncertainty relation between any pair of "conjugate" quantities. The momentum and position are conjugate, and Heisenberg's commutator can be realized by the representation $p = -i\hbar\frac{\partial}{\partial x}$. This says that momentum is the generator of spatial displacement.

Similarly, Schrödinger's equation represents the energy with $i\hbar\frac{\partial}{\partial t}$, making it the generator of time evolution. Thus we have the energy uncertainty relation

$$\Delta E \Delta t > \hbar.$$

According to this relation, a state with definite energy ($\Delta E = 0$), will last indefinitely ($\Delta t = \infty$). One with uncertain energy, called a *virtual state*, has a limited lifetime $\hbar/\Delta E$. Experimentally we have observed unstable particles with lifetimes ranging from 10^{-23} s to thousands of years.

There is, however, no sharp dividing line between stable and unstable states. A state with a lifetime of a hundred years will appear to us as stable for all practical purposes. Particles that we think would live forever, such as the proton, may well have unknown interactions that give it a very long but finite lifetime.

In 1948, Richard Feynman give a reformulation of quantum theory that brings out the fact that virtual states include anything you can think of, and more.

Quantum mechanical processes are described though transition amplitudes between states. The probability of a transition is the squared modulus of the corresponding amplitude. If we have a way to calculate the amplitude for all conceivable processes, that defines

---

[3]剎那: in Zen, instant of time.

Richard P. Feynman (1918–1988)

Fig. 6.1 After twenty years, a new formulation of quantum mechanics.

the theory. Feynman gave a formula for that amplitude:

$$\text{Transition amplitude} = \sum_{\text{history}} \exp \frac{i}{\hbar} \left(\text{Action of a history} \right),$$

where "action of a history" refers to the action of a classical path connecting the initial state to the final state. The sum is to be carried out over all possible paths. Since the paths form a continuous set, the sum is actually an integral. It is called the *Feynman path integral*.

As we can see, the Feynman amplitude is a sum of phase factors proportional to $\hbar^{-1}$. The limit $\hbar \to 0$ corresponds to classical physics. In this limit, any small variation in the action will be infinitely magnified. The phase angle will go through a large number of 2π rotations, and become essentially random. Thus, contributions of different histories will tend to cancel each other, leaving only the contribution of the history that minimizes the classical action, and we have classical physics.

For finite $\hbar$, all histories contribute, yielding quantum corrections to classical physics. Since $\hbar$ has dimension of (energy × time),
whether it should be considered large or small depends on characteristic parameters of this dimension associated with the initial and
final states. If $\hbar$ is effectively small, then the transition goes through
classically, with small quantum fluctuations. Otherwise non-classical
paths will be important. "Outlandish" histories may have a relatively
large action, and thus make small contributions to the transition amplitude, but they are present.

A history is a "virtual reality". According to the Feynman path
integral, you can construct a quantum system by choosing an allowed
set of virtual realities, with specified classical actions. In this manner, you can unleash your imagination in ways not accessible in the
canonical formulation of quantum mechanics. For example, you could
allow space-time to have any number of dimensions, or to have any
form of curvature. In order to do this, you must be able to write down
a meaningful action covering these possibilities. You then sum over
all possible dimensions, or all possible metric functions. The least
action will pick out the correct dimension, or metric, in the classical
limit of your theory.

Perhaps, in this manner, we might someday find answers to "deep"
questions, such as why space-time appears to have four dimensions.

6.5. The quantum century

Quantum mechanics burst upon the 20th century and made it her
own. Technology had advanced to such a degree that inventions
sparked by pure science rode a very short fuse. In three quarters
of a century, quantum mechanics gave birth to fields that took the
world by storm. Among these are:

- **atomic and molecular physics**, which finally and firmly made
  chemistry a deductive science;
- **nuclear physics**, which led to the technological and political up-
  heavals associated with the name "nuclear age";

- **solid-state physics**, which gave us, among other things, the computer chip and information technology.

All that, because we have mastered the fact that position and momentum do not commute.

6.6. *The Waste Lecture*

Excerpts from a poem entitled *The Waste Lecture* attributed to T. S. Eliot (1888–1965):[4]

> Momentum is not well defined, being
> Canonical to place, failing
> To commute exactly, leaving
> Necessary doubt.
> Newton spoke firmly, writing
> Definitive equations, moving
> His particles on clean trajectories.
>
> And when we were pupils, studying the rudiments,
> How confident we were, precisely calculating
> x and p (not one but both!) with such abandon.
> But at the university our teachers —
> Murmuring of commutation — frowned and flunked us.
> We read, much of the night, but are none the wiser.

---

[4]*Physics Today*, April 1 (1989), with comments from John Lowell of Manchester, England that the poem contains "unmistakable echoes of the *Wasteland*", and was "strongly influenced by the quantum theory that was growing vigorously when Eliot was a young man".

7

What is Charge?

7.1. The quantum gauge

In classical theory, the vector potential of the electromagnetic field can freely undergo gauge transformations. The "gauge" has no impact on the physics, because it does not alter the electric and magnetic fields. Since classical charged particles interact with the electric and magnetic fields, they never directly see the vector potential, and have no knowledge of the gauge.

In contrast, a charged particle in quantum theory interacts with the vector potential, as we shall explain later. It knows about the gauge, and must act in such a manner as to preserve the physics.

A gauge transformation in quantum theory involves both the vector potential and the charged particle. It consists of the joint operation

$$A \to A + \partial\chi, \qquad \psi \to U\psi,$$

where A is the vector potential, ψ is the particle's wave function, and U is a phase factor:

$$U = \exp\left(\frac{iq}{\hbar c}\chi\right).$$

Here, q is the charge of the particle.

In the fiber bundle representing the vector potential, we must now associate a ring with each fiber in order to register the quantum phase of the particle, as illustrated in Fig. 7.1. When the vector

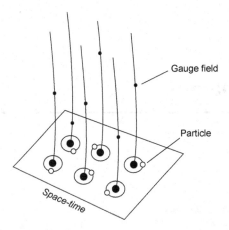

Fig. 7.1   The fiber of the vector potential is augmented by a ring, representing the quantum phase of the particle, which changes during a gauge transformation. The charge is the generator of the gauge transformation. Compare with Fig. 4.1.

potential climbs up and down a gauge fiber, the phase makes correlated rotations around the ring. This happens independently at all space-time points.

Note that the quantum phase is proportional to the charge of the particle. A neutral particle with $q = 0$ will not take part in the gauge transformation. Herein lies the fundamental definition of charge:

Charge is the generator of gauge transformations.

7.2. Covariant derivative

Why does the quantum phase change in a gauge transformation?

In order to turn on the electromagnetic coupling, we make the replacement $p \to p - \frac{q}{c}A$, according to the gauge principle. In classical mechanics, this is just an numerical substitution; but in quantum mechanics p is an operator represented by $p = i\hbar\partial$. This means that, in the Schrödinger equation, we make the substitution $\partial \to D$, with

$$D = \partial + \frac{iq}{\hbar c}A \, .$$

David Bohm (1917–1992)

Fig. 7.2 "Charged particles are directly coupled to a gauge field" — result of experiment to prove quantum mechanics wrong, but reaffirmed it instead.

This is called a "covariant derivative".

The Schrödinger equation is gauge invariant because the term arising from A is cancelled through the action of ∂ on the phase factor U in the wave function.

7.3. Aharonov–Bohm experiment

The vector potential was optional in classical electromagnetism but is mandatory in quantum mechanics, because it appears explicitly in the equation of motion. On the other hand, it is not directly observable, being determined only up to a gauge transformation.

David Bohm found this situation curious. He sided with Einstein in the belief that quantum mechanics was "incomplete", and thought that the strange role of the vector potential was a fatal flaw in the theory. In 1959 he and Y. Aharonov proposed an experiment to test this hunch.

When an electric current flows through a solenoid, it creates a magnetic field largely confined within the solenoid, except for fringing

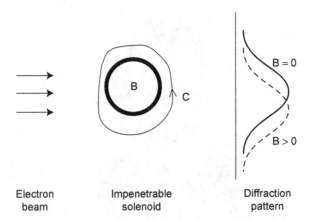

Electron   Impenetrable   Diffraction
beam   solenoid   pattern

Fig. 7.3   The Aharonov–Bohm experiment demonstrates the reality of the gauge field. Electrons are scattered by an impenetrable solenoid, in which a magnetic field can be turned on and off. When the field is turned on, the diffraction pattern shifts, even though electrons never feel the magnetic field. This is because an electron is coupled to the vector potential — the gauge field — which is not zero outside the solenoid. Its line integral over a closed path C is equal to the magnetic flux inside the solenoid.

effects. The leakage to the outside can be made minimal by making the solenoid very long. The vector potential, however, cannot be zero outside the solenoid, because according to Maxwell's equations the line integral

$$\oint \mathbf{A} \cdot d\mathbf{x}$$

along a closed circuit around the solenoid must equal the magnetic flux inside. This is illustrated in Fig. 7.3, with the solenoid shown in cross section.

   If the solenoid is made impenetrable to an electron, common sense suggests that the electron does not know about the magnetic field inside. But quantum mechanics says it knows, because it can feel the vector potential outside.

   In the proposed experiment, an electron beam is scattered by an impenetrable solenoid, and forms a diffraction pattern on an observation screen downstream. Classical reasoning says that the diffraction pattern should be the same whether or not there is a magnetic field inside the solenoid. Quantum theory, however, predicts that the diffraction pattern will be shifted when the magnetic field is turned on.

The experiment was performed, and the diffraction pattern did shift, precisely as predicted by quantum mechanics.[1] This affirms the gauge principle in quantum theory, and the fundamental role of the vector potential. Note that the vector potential remains unobserved, because the experiment measures not A but the line integral $\oint \mathbf{A} \cdot d\mathbf{x}$, which depends only on the magnetic flux inside the solenoid.

7.4. $U(1)$

The quantum phase factor

$$\exp\left(-\frac{iq}{\hbar c}\chi\right)$$

is a "unitary" operation on the wave function, in the sense that it does not affect the modulus of the wave function. It is a representation of the mathematical group $U(1)$ (unitary group of dimension 1), the group of all rotations about a fixed axis.

The phase factor exhibits periodic behavior, in that it returns to an original value whenever the phase increases by 2π. Thus, while the phase angle can go from 0 to ∞, the phase factor has only a finite range of values. Mathematically we say that it is a "compact" representation of $U(1)$.

Suppose there are two species of particles of charges q and q'. The fact that they are coupled to one universal electromagnetic field means that, under a gauge transformation, their wave functions undergo phase changes proportionate to the respective charges. We can

---

[1]The experiment was proposed in Y. Aharonov and D. Bohm, *Phys. Rev.* **115**, 485 (1959). Reliable experimental verification came more than twenty years later, in N. T. Osakabe *et al.*, *Phys. Rev. A* **34**(2), 815 (1986).

represent the particles by a two-component wave function:

$$\Psi = \begin{pmatrix} \psi \\ \psi' \end{pmatrix}.$$

Then the gauge transformation is represented by the 2×2 matrix

$$\begin{pmatrix} \exp\left(-\dfrac{iq}{\hbar c}\chi\right) & 0 \\ 0 & \exp\left(-\dfrac{iq'}{\hbar c}\chi\right) \end{pmatrix}.$$

This is a two-dimensional representation of $U(1)$. As χ varies, the matrix generally does not repeat itself, and the range of values of the matrix is unbounded. However, if charge is quantized, i.e. if q and q' are integer multiples of a basic unit, then the matrix will repeat itself as χ increases, making the representation compact.

Thus, requiring that representations of $U(1)$ be compact leads to *charge quantization*. This is one way to obtain this condition. Another way is to have a Dirac monopole, which will be discussed in Chap. 9.

7.5. Quantum gauge principle

The gauge principle can be stated in an alternative form, from the point of view of gauge symmetry.

Not all systems can be coupled to the electromagnetic field. Those that can must possess *global gauge invariance* before the coupling is turned on. This means that the Schrödinger equation should be invariant under a constant phase change:

$$\psi \to e^{i\alpha}\psi\,,$$

where α is a constant. The usual form of the Schrödinger equation has such invariance, because the constant phase factor "slips through" a differentiation: $\partial(e^{i\alpha}\psi) = e^{i\alpha}(\partial\psi)$. This global gauge invariance guarantees the existence of a conserved current that expresses charge conservation.

Without electromagnetic coupling, the system is not invariant under a *local gauge transformation* — one with a phase that depends on space-time:

$$\psi \to e^{i\beta(x)}\psi.$$

The original Schrödinger equation is not invariant under this transformation, because the phase factor can now no longer escape differentiation.

To make the system invariant under a local gauge transformation, we make the replacement $\partial \to D$. The term generated by differentiation of the phase factor is now cancelled by that arising from the vector potential.

To reiterate:

• First consider a matter system with global gauge invariance, which guarantees existence of a conserved charge.

• Extend the global gauge invariance to local gauge invariance, through the replacement $\partial \to D$, thereby introducing coupling to a gauge field.

In short, the quantum gauge principle states that:

Coupling to a gauge field promotes global gauge invariance to local gauge invariance.

7.6. Global vs. local gauge invariance

The difference between a global and a local gauge transformation may be illustrated in Fig. 7.4. In a global gauge transformation, the quantum phase runs around the rings in the same manner over all space-time. In a local gauge transformation, it changes independently at each space-time point, but the change must be correlated with a movement of the gauge field along its fiber.

If we only had global gauge invariance, then, while a charge can be regarded as positive or negative as a matter of definition, the same definition must be used throughout space-time.

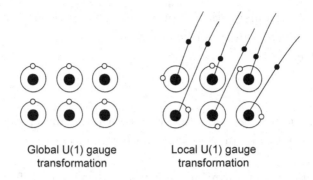

Global U(1) gauge   Local U(1) gauge
  transformation   transformation

Fig. 7.4 Global and local gauge transformations. Left panel: in a global gauge transformation, the quantum phase change is the same at all space-time points. Right panel: in a local gauge transformation, the quantum phase can have arbitrary independent values at different space-time points, but it is correlated with the gauge field, whose gauge function (position on its fiber) keeps track of the phase.

For example, suppose there were no electromagnetic coupling. Then, we are free to call the electron charge negative or positive on Earth, but the same convention must be adhered to on Mars.

With local gauge invariance, which requires the presence of a gauge field, the convention for charge becomes a purely local matter. An observer on Mars can define an electron as positive, while on Earth we continue to regard it as negative. When the Mars observer sends us an electron, it interacts with ours correctly, because the interaction occurs through the gauge field, which keeps track of the local protocols.

Local gauge invariance frees us from the last vestige of "action at a distance".

8

The Zen of Rotation

Maxwell's gauge theory is an expression of local gauge invariance. It enables the charged particle to change its quantum phase freely and independently at different points of space-time. Mathematically, a phase change is equivalent to a rotation about a fixed axis. The central theme of this book, the Yang–Mills gauge theory, lifts the restriction to a fixed axis. To appreciate that, we have to first understand the essence of rotation.

8.1. Rotations do not commute

When we make two successive rotations about a fixed axis, the order of operation makes no difference in the outcome. That is, elements of the group $U(1)$ commute with one another. However, rotations about different axes do not commute with each other. A demonstration of this fact is depicted in Fig. 8.1, in which a book is being rotated 90° successively about a horizontal axis and a vertical axis. As we can see, the outcome depends on the order of the operations.

A $U(1)$ rotation through angle θ can be represented by the phase factor $e^{i\theta}$. If the angle is infinitesimally small, this reduces to

$$1 + i\theta \,.$$

Any rotation about a fixed axis can be composed from successive infinitesimal rotations about that axis.

A general rotation can be carried out about any one of three independent axes in space, and consequently there are three possible

Fig. 8.1 Rotations do not commute: A book is rotated successively about a horizontal axis and a vertical axis, in different orders, with different results. Top row: 90° about horizontal axis, then 90° about vertical axis. Bottom row: same rotations in reversed order.

infinitesimal rotations, which we represent as

$$1 + i\theta_1 L_1\,,$$
$$1 + i\theta_2 L_2\,,$$
$$1 + i\theta_3 L_3\,.$$

We have to insert the quantities L_1, L_2, L_3 to make these operations non-commutative. These are operators called "generators" of rotation. Any rotation can be composed of factors like these, but the order of the factors is important.

The mathematical structure of the group of rotations is completely specified by the commutator $[L_a, L_b] = L_a L_b - L_b L_a$, for all pairs of generators.

8.2. Hamilton's flash of insight

William Rowan Hamilton, the Irish mathematician who gave us the Hamiltonian, had thought long and hard about rotations. He referred to the generators as a "triplet", and denoted them i, j, k. These form

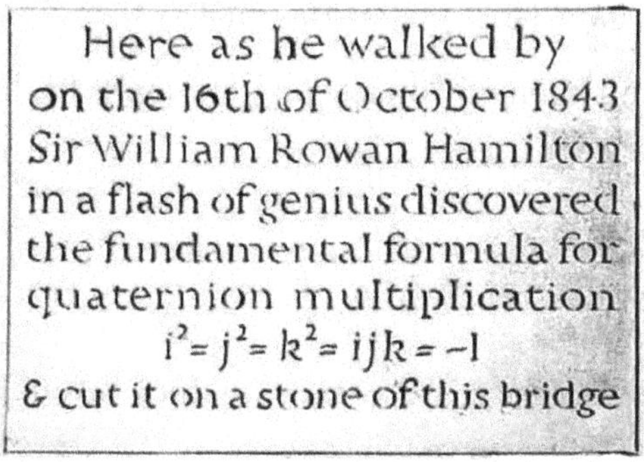

Here as he walked by
on the 16th of October 1843
Sir William Rowan Hamilton
in a flash of genius discovered
the fundamental formula for
quaternion multiplication
$i^2 = j^2 = k^2 = ijk = -1$
& cut it on a stone of this bridge

Fig. 8.2 Plaque on Brougham Bridge, Dublin, Ireland, commemorating the moment Hamilton perceived the essence of rotation.

a quartet together with the identity. Through superposition, one can generate "quaternions", which Hamilton regarded as extensions of complex numbers.

Hamilton knew how to add triplets, but not multiply them, and this had caused great frustration.

One day in 1843, as Hamilton was walking on the Brougham Bridge in Dublin, the rule of triplet multiplication came to him in a flash, and he scratched it down on the bridge with his knife:

$$i^2 = j^2 = k^2 = ijk = -1\,.$$

The scratch marks have long been eroded, but a plague commemorating the event stands, as pictured in Fig. 8.2.

Some twenty years later, Hamilton recalled the discovery in a letter to his son:[1]

> Every morning, on my coming down to breakfast, your (then) little brother William Edwin, and yourself, used to ask me, "Well, Papa, can you multiply triplets?" Whereto I was always obliged to reply, with a sad shake of the head: "No, I can only add and subtract them."

[1]From letter of William Rowan Hamilton to the Rev. Archibald H. Hamilton, dated August 5, 1865.

But on the Brougham Bridge that day in 1843,

An electric circuit seemed to close; and a spark flashed forth, ...

and he knew how to multiply triplets.

8.3. Generators of rotation

Hamilton's triplet is related to our generators through the identification

$$i = -2\iota L_1, \qquad j = -2\iota L_2, \qquad k = -2\iota L_3,$$

where we have used ι temporarily to denote the pure imaginary $\sqrt{-1}$. Hamilton's formulas translate to the commutator

$$[L_a, L_b] = i\epsilon_{abc}L_c.$$

(We have restored i as the pure imaginary.) The repeated index c is summed over. The symbol ϵ_{abc} is the "completely antisymmetric tensor of rank 3".[2] In a single formula, the above summarizes the following properties:

$$L_1^2 = L_2^2 = L_3^2 = \frac{1}{4},$$
$$L_1 L_2 = -L_2 L_1 \text{ etc.,}$$
$$L_1 L_2 = \frac{i}{2} L_3 \quad \text{(and cyclic permutations).}$$

8.4. Groups

Some terminology relevant to gauge theories.

A mathematical group is a set of things for which a binary relation is defined, called "group multiplication" so that the product of two elements of the group is also an element of the group. Also, it

---

[2] The possible values of ϵ_{abc} are $0, 1, -1$. It is 0 if any two indices are the same. Thus it is non-zero only when $\{a, b, c\}$ is $\{1, 2, 3\}$, in some order. It is 1 if the order is a cyclic permutation of 123, otherwise it is -1.

Neils Henrik Abel   Marius Sophus Lie
(1802–1829)   (1842–1899)

Fig. 8.3 Two Norwegian mathematicians who leave marks on gauge theory, which relies on non-Abelian Lie groups.

should contain an identity element, and every element should have an inverse, whose product with the element gives the identity.

For example, the set of all positive and negative integers form a discrete group, with addition as group multiplication, 0 as the identity, and the inverse of an element its negative. In this example, the group multiplication is commutative, and such groups are called *Abelian*, after the Norwegian mathematician Niels Henrik Abel.

Groups with non-commuting multiplication are called *non-Abelian*. Thus, the rotation group is non-Abelian, while the $U(1)$ subgroup of rotations about a fixed axis is Abelian.

Groups with continuous elements are called Lie groups after another Norwegian mathematician Sophus Lie. The generators are defined by commutation relations of the form

$$[L_a, L_b] = iC^c_{ab}L_c \,,$$

where the coefficients C^c_{ab} are called structure constants. They form a closed set under commutation called the "Lie algebra" of the group. For the rotation group, $C^c_{ab} = \epsilon_{abc}$.

8.5. $SU(2)$: fundamental representation

The generators L_a cannot be numbers, since they do not commute with each other. We may represent them by matrices, and the most economical representation uses 2×2 matrices. A standard set are the Pauli matrices:

$$\sigma_1 = \begin{pmatrix} 0 & 1 \\ 1 & 0 \end{pmatrix}, \quad \sigma_2 = \begin{pmatrix} 0 & -i \\ i & 0 \end{pmatrix}, \quad \sigma_3 = \begin{pmatrix} 1 & 0 \\ 0 & -1 \end{pmatrix}.$$

Any 2×2 matrix can be written as a linear combination of these, together with the identity matrix. The generators are represented by $L_a = \frac{1}{2}\sigma_a$.

The most general rotation can be written in the form

$$U = \exp\left(\frac{i}{2}\omega_a\sigma_a\right),$$

where ω_a are real numbers.[3] This is a 2×2 unitary matrix with unit determinant, and they form the group $SU(2)$ — special unitary group of dimension 2. More precisely, these matrices furnish the *fundamental representation* of $SU(2)$, the faithful representation of the smallest possible dimension. There are representations with higher dimensionality, in particular the 3-dimensional adjoint representation introduced below.

The Pauli matrices correspond to Hamilton's triplet, apart from a factor $\sqrt{-1}$. The 2×2 matrices forming the fundamental representation of $SU(2)$ are equivalent to Hamilton's quaternions.

If a physical system has $SU(2)$ internal symmetry, the fundamental representation is realized by two-component wave functions:

$$\psi = \begin{pmatrix} \psi_1 \\ \psi_2 \end{pmatrix}.$$

These describe a spin 1/2 particle, whose probability amplitude for up and down spin are respectively ψ_1 and ψ_2. A rotation of the system in internal space is represented by $\psi \to U\psi$.

---

[3]The exponential function is defined by its power series, $\exp z = 1 + z + \frac{1}{2}z^2 + \frac{1}{2\cdot3}z^3 + \cdots$, where z can be anything. Thus, the exponential of an $n \times n$ matrix is an $n \times n$ matrix.

8.6. The adjoint representation

With any Lie group comes a natural representation called the adjoint representation. Built directly from the structure constants, it consists of matrices of dimensionality equal to the number of generators. In this representation the generators are given by

$$(L_a)_{bc} = -iC^c_{ab}.$$

The very naturalness of this representation gives it a special role in gauge theories.

As an illustration, the adjoint representation of $SU(2)$ consists of the 3×3 matrices $L_a = -i\epsilon_{abc}$:

$$L_1 = \begin{pmatrix} 0 & 0 & 0 \\ 0 & 0 & -i \\ 0 & i & 0 \end{pmatrix}, \quad L_2 = \begin{pmatrix} 0 & 0 & i \\ 0 & 0 & 0 \\ -i & 0 & 0 \end{pmatrix}, \quad L_3 = \begin{pmatrix} 0 & -i & 0 \\ i & 0 & 0 \\ 0 & 0 & 0 \end{pmatrix}.$$

This answers a burning question:

Q. Rotations in 3D form a group $SU(2)$. What happens to the "3" in 3D?

A. It's the number of generators, the dimension of the adjoint representation. The "2" in $SU(2)$ is the dimension of the fundamental representation.

The relation between the fundamental and adjoint representation is rather intriguing, but we will not digress.[4]

---

[4]For explanation and a graphic demonstration between the fundamental and adjoint representations of the rotation group, see K. Huang, *Quarks, Leptons, and Gauge Fields*, 2nd edn. (World Scientific, Singapore, 1992), pp. 65–66.

9

Yang–Mills Field: Non-Commuting Charges

9.1. Gauging $SU(2)$

The electromagnetic interaction is created by "gauging $U(1)$".

This means that we start with a particle system that has global gauge invariance under the group $U(1)$, and then extend the symmetry to local gauge invariance by coupling the particle to a gauge field — the vector potential.

Chen-Ning Yang and Robert L. Mills[1] generalized this procedure to the non-Abelian group $SU(2)$. This results in an extension of Maxwell's equations to new types of interactions, which turn out to cover all fundamental interactions among elementary particles.

Yang and Mills were motivated by the conservation of isotopic spin, an attribute like spin. The proton and the neutron can be regarded as the "up" and "down" states of the nucleon, a particle with isotopic spin $1/2$.[2] The strong nuclear force treats proton and neutron on the same footing. This corresponds to a global gauge invariance under $SU(2)$. Yang and Mills states:

> The conservation of isotopic spin is identical with the requirement of invariance of all interactions under isotopic spin rotation. This means that the orientation of the isotopic spin is

---

[1] C. N. Yang and R. L. Mills, *Phys. Rev.* **96**, 191 (1954).

[2] In modern usage, isotopic spin is shortened to *isospin*. The name is derived from "isotope", which means same number of protons, but different number of neutrons.

Chen-Ning-Yang (1922–)

Fig. 9.1   From Maxwell to non-Abelian gauge theory.

of no physical significance. The differentiation between a neutron and a proton is then a purely arbitrary process. As usually conceived, however, once one chooses what to call a proton, what to call a neutron, at one space-time point, one is then not free to make any choices at other space-time points.

It seems that this not consistent with the localized field concept that underlies the usual physical theories.

Mathematically, we describe the nucleon by a two-component wave function:

$$\psi = \begin{pmatrix} \text{Proton} \\ \text{Neutron} \end{pmatrix}.$$

An isotopic-spin rotation is represented by $\psi \to U\psi$, where U is a 2×2 matrix belonging to the fundamental representation of $SU(2)$. Global gauge invariance means that the equation of motion should be invariant when U is a constant matrix.

In infinitesimal form, we have $U = 1 + iL_a\omega_a$, where L_a are the three generators of $SU(2)$, and ω_a are infinitesimal constant

Fig. 9.2 Chen-Ning Yang and Robert L. Mills (1927–1999), at symposium honoring Yang's retirement, State University of New York at Stony Brook, 1999.

parameters. As a consequence of global gauge invariance, there are three conserved isotopic spin currents j_a, such that $\partial \cdot j_a = 0$.

To gauge $SU(2)$, we do the following:

- introduce a 4-vector gauge field A_a with three internal components labeled by $a = 1, 2, 3$, corresponding to the three generators of the gauge group;
- replace the derivative ∂ by the covariant derivative

$$D = \partial + \frac{ig}{\hbar c} L_a A_a \,.$$

In the equation of motion, this covariant derivative generates a coupling between the particle and the gauge field, with interaction energy density

$$j_a \cdot A_a \,,$$

where j_a is the conserved isotopic spin current density.

The nucleon is now endowed with three isotopic charges $g L_a$, where g is the gauge coupling constant. The novel feature is that the charges do not commute with one another.

9.2. Picturing local gauge invariance

The system is now invariant under a local gauge transformation.

For simplicity we shall use the shorthand $A = A_a L_a$, which is a matrix. An infinitesimal local gauge transformation can be written in the form

$$\psi \to (1 + ig\chi)\,\psi\,,$$
$$A \to A + \partial\chi - ig[\chi, A]\,,$$

where χ_a are arbitrary infinitesimal functions of space-time.

Comparing the transformation to that in the Abelian case, $A \to A + \partial\chi$, we see that there is an extra term $-ig\,[\chi, A]$. This describes a mixing of components of gauge field according to the adjoint representation of $SU(2)$.

To visualize the gauge transformation, imagine a gyroscope attached to each point of space-time, as schematically depicted in Fig. 9.3. A global gauge transformation is a rotation of all the gyroscopes in unison. Introduction of the gauge field attaches a fiber to each space-time point, with three beads moving along each fiber representing the components of the gauge field. In a local gauge transformation, the gyroscopes rotate independently, while the beads slide on its fiber in correlated moves.

Yang and Mills used isotopic spin to illustrate a principle. In the real world, isotopic spin is not conserved, being violated by the electromagnetic interaction. The violation is relatively weak; and the proton and neutron masses are only slightly different. But no matter how *petit* the difference, the particles are distinct and cannot be mixed.

Only *exact* symmetries can be gauged.

9.3. Maxwell generalized

In Maxwell's $U(1)$ gauge theory, we define a gauge-invariant field tensor $F^{\mu\nu} = \partial^\mu A^\nu - \partial^\nu A^\mu$, whose components are the electric and magnetic fields. In the non-Abelian case, however, such a tensor is

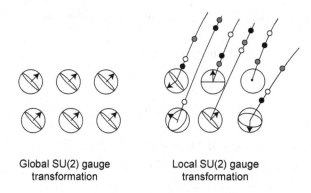

Global SU(2) gauge   Local SU(2) gauge
transformation   transformation

Fig. 9.3 In a global $SU(2)$ gauge transformation, symbolic gyroscopes attached to points of space-time rotate in unision. In a local transformation, they rotate independently, but three gauge fields undergo correlated gauge transformations. The latter is indicated by the positions of three beads on a fiber. Compare with Fig. 7.4.

not gauge invariant. In fact, there is no gauge-invariant field tensor.

The next best thing is to consider the "gauge covariant" quantity

$$F^{\mu\nu} = \partial^\mu A^\mu - \partial^\nu A^\mu + ig\,[A^\mu, A^\mu],$$

where $F^{\mu\nu} = F_a^{\mu\nu} L_a$. By gauge covariant we mean that it transforms according to the adjoint representation of the gauge group. This turns out to be the correct choice for field tensor.

Yang had searched for this tensor without success since his student days in 1947. As he recalls:[3]

> I was clearly focusing on a very important problem. Unfortunately the mathematical calculations always ended in more and more complicated formulas and total frustration. It was only in 1953–1954, when Bob Mills and I revisited the problem and tried adding quadratic terms to the field strength $F^{\mu\nu}$ that an elegant theory emerged. For Mills and me it was many years later that we realized the quadratic terms were in fact natural from the mathematical point of view.

[3]C. N. Yang, in *50 Years of Yang–Mills Theory*, ed. G. 't Hooft (World Scientific, Singapore, 2005), p. 7.

The "natural mathematical point of view" refers to the adjoint representation.

The equations of motion turn out to be

$$D_\mu F^{\mu\nu} = -j^\nu$$
$$D_\mu \tilde{F}^{\mu\nu} = 0 .$$

These are generalizations of Maxwell's equations (Sec. 3.9).

The first of these equations can be rewritten

$$\partial_\mu F^{\mu\nu} = -j^\nu - ig[A_\mu, F^{\mu\nu}] .$$

The right-hand side gives the current that generates the field. Note that it contains the gauge field itself. That is, the gauge field carries charge, and acts as its own source. In contrast, the electromagnetic field is neutral, and does not have intrinsic self-interaction.

9.4. Gauge photons

The equation of motion for weak fields reduces to the form

$$\Box A = 0 ,$$

where non-linear terms have been neglected. This says that quantization of the theory will yield massless gauge photons which, like ordinary photons, are spin 1 bosons. The important difference is that the gauge photons in this case carry charge, and consequently have intrinsic interactions with each other.

One way to give mass to the gauge photon is to modify the linearized field equation to read

$$\Box A + \left(\frac{m}{\hbar c}\right)^2 A = 0 ,$$

where m is the mass. This is not acceptable, however, since the added term destroys gauge invariance.

Gauge invariance guarantees that the gauge photons are massless.

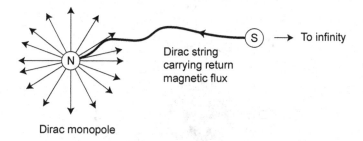

Dirac monopole

Fig. 9.4 Dirac isolates a magnetic north pole from a dipole, by sending the south pole to infinity, and concentrating the flux between them in a very thin string. The string becomes invisible to particles when a charge-quantization condition is satisfied.

9.5. Magnetic charge

The analog of the magnetic field in Yang–Mills theory is

$$\mathbf{B}_a = \nabla \times \mathbf{A}_a + \frac{1}{2}g\epsilon_{abc}\mathbf{A}_b \times \mathbf{A}_c.$$

Thus,

$$\nabla \cdot \mathbf{B}_a = \frac{1}{2}g\epsilon_{abc}\nabla \cdot (\mathbf{A}_b \times \mathbf{A}_c),$$

which is the content of the second set of the equations of motion.

In the Maxwell case, we had $\mathbf{B} = \nabla \times \mathbf{A}$, and $\nabla \cdot \mathbf{B} = 0$, and the last relation indicates the absence of magnetic charge. Now, there is magnetic charge density arising from the self-interaction of the gauge field.

9.6. Monopole: the gauge hedgehog

Since there is magnetic charge density, we should be able to build a magnetic monopole from Yang–Mills fields.

Actually, a magnetic monopole can exist in Abelian theory, albeit "with strings attached". This is the "Dirac monopole" depicted in Fig. 9.4. You start with a magnetic dipole, send the south pole to infinity, and squeeze the magnetic flux between the poles into a thin

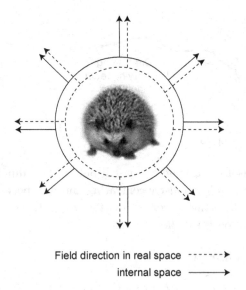

Field direction in real space ------►
internal space ———►

Fig. 9.5   Gauge hedgehog: Yang–Mills monopole with no strings attached. The magnetic field is a vector both in 3D space, as well as in 3D internal symmery space. In "Coulomb gauge" the internal vector points radially outward, like the quills of a hedgehog.

string.[4] This "Dirac string" can be made invisible to a charged particle, if the strength g of the magnetic monopole obeys the condition $ge = n/2$, where e is the charge of the particle, and n is an integer. Thus, the mere possibility that a Dirac monopole exists implies charge quantization: $e = n/2g$.

In Yang–Mills theory, one can construct a field configuration corresponding to a monopole without the Dirac string. It involves an interesting correlation between the orientation of the gauge field in internal symmetry space and in ordinary space. In the "Coulomb gauge" $\nabla \cdot \mathbf{A}_a = 0$, the field in internal space points along the radial direction in ordinary space, as depicted in Fig. 9.5. This field configuration is called a "gauge hedgehog", and was first constructed by Yang and Tai-Tsun Wu.

[4]P. A. M. Dirac, *Proc. Roy. Soc. London*, **133**, 60 (1933).

Fig. 9.6 C. N. Yang with Tai-Tsun Wu (left panel) and Gerald 't Hooft (right panel), two contributors to the theory of the monopole in Yang–Mills theory.

In pure Yang–Mills theory, the hedgehog has infinite energy, owing to a singularity at the origin. Years later, in the context of a massive theory with spontaneous symmetry breaking, Gerald 't Hooft constructed a monopole with finite energy.

9.7. Into the deep freeze

The gauge photons in Yang–Mills theory are massless by virtue of gauge invariance. This poses an obstacle to physical applications, because we know of no massless vector particles besides the photon. For this reason, the Yang–Mills theory promptly went into hibernation, while the physics world swept by with sound and fury, driven by the boom in particle accelerators.

One day, it will be resurrected, but that moment lies more than a decade ahead (Chap. 17).

When push comes to shove, the mass problem will be overcome through spontaneous symmetry breaking (Chap. 18).

10

Photons Real and Virtual

10.1. Real photons

In the brave new world of quantum theory, even as matter becomes waves, light becomes particles.

Planck and Einstein suggested, in 1900 and 1905 respectively, that light comes in discrete packets of energy called photons. Planck based his view on a study of the spectrum of radiation emitted by hot bodies, while Einstein postulated the photon to explain the experimental photoelectric effect. From different avenues, they arrive at the same conclusion, that a photon of frequency ν has energy

$$E = h\nu \,,$$

where h is Planck's constant.

However, light is not composed of particles in the classical sense, just as matter is not your ordinary wave. The words are metaphors that should not be taken literally.

In the macroscopic world, you can create radiation by waving a charge back and forth. The disturbance in the electromagnetic field propagates at a finite speed, and soon embarks on its own journey as free radiation. Solving Maxwell's equations with appropriate initial conditions, in principle, yields the complete history of the electromagnetic field everywhere.

In the microscopic world, elementary particles radiate by emitting photons, one at a time. They can also absorb photons, one at a time. These processes are quantum transitions described by transition amplitudes, whose squared modulus gives the transition rate,

Fig. 10.1 Max Planck (1858–1947) on the German two-mark coin. The constant named after him sets the scale of quantum phenomena.

which is what experiments measure, and what we want to calculate theoretically.

A photon is an elementary massless particle of spin 1 (in units of $\hbar$). It always moves at the velocity of light, with momentum $p = E/c$. As illustrated in Fig. 10.2, the spin may be either parallel or antiparallel to the momentum, corresponding to left and right circular polarizations.[1] The momentum and the polarization are the "quantum numbers" that specify the state of a photon.

When one mole of hydrogen gas is stimulated to emit light, the 10^{23} electrons in the 10^{23} hydrogen atoms of the gas emit photons, one at a time (generally not in unison). They produce a dense photon gas that can be described by the classical electromagnetic field.

Although photons are created one at a time, successively created photons are correlated, and exhibit diffraction phenomena. When impinging on two slits, photons go through either one of the slits, one at a time, but they create a diffraction pattern on a detection screen behind the slits.

The earliest demonstration of photon interference was perhaps G. I. Taylor's 1909 experiment,[2] motivated by J. J. Thomson's doubt

---

[1] Right circular polarization means the electric field rotates to the right when you look at the photon head-on. This corresponds to the spin pointing opposite the momentum.
[2] G. I. Taylor, "Interference fringes with feeble light", *Proc. Camb. Phil. Soc.* **15**, 114 (1909).

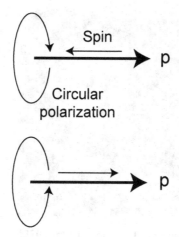

Fig. 10.2 A photon of given momentum has two spin states, corresponding to right and left circular polarizations.

that interference can be observed for extremely weak light. Taylor successfully photographed the diffraction pattern of a needle, at a luminosity equivalent to "a candle burning at a distance of slightly less than one mile." The exposure time was three months, during which time he went on a yacht trip.

The output from a candle at one mile is about one million photons of yellow light, per second, per square inch. This works out to one photon every two seconds striking a photographic grain of size 10 microns.

10.2. Quantum jumps

In quantum theory, the vacuum is filled with fluctuating electric and magnetic fields, and an electron is being buffeted as if in a stormy sea. If the electron was in an excited orbit of an atom, this perturbation will induce it to jump to an orbit with lower energy, while emitting a photon. This is illustrated in Fig. 10.4.

The quantum states of the electromagnetic field, and those of the electron, belong to separate spaces. The quantum jump occurs between two states of the joint field–particle system, with the following

Geoffrey I. Taylor (1886–1975)

Fig. 10.3 Imaging diffraction pattern from a candle a mile away, by collecting one photon every two seconds, for three months.

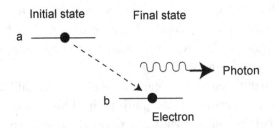

Fig. 10.4 A particle makes a quantum jump from state a to state b, and emits a photon. The transition amplitude is the matrix element of the interaction Hamiltonian between the intial and final states.

initial and final states:

Initial state = (Photon vacuum state) × (Electron state a),

Final state = (One photon state) × (Electron state b).

The probability amplitude for the transition is given by a "transition matrix element", which can be calculated from the interaction energy (interaction Hamiltonian). The square modulus of the amplitude gives the transition probability.

Experimentally we observe a steady-state process in which transitions occur continually, and we measure the transition rate, which is given by *Fermi's golden rule*:

$$\text{Transition rate} = \frac{2\pi\rho}{\hbar}|\text{Transition matrix element}|^2 ,$$

where ρ is the density of final photon states at the given energy. The inverse of the transition rate gives the mean lifetime of the initial electron state.

Just as an electron can emit a photon, it can absorb one. To calculate the absorption rate, all we have to do to is to reverse the initial and final states.

The initial electron could decay via other modes, for example by emitting more than one photon. Indeed, it can do anything allowed by the interactions, and by energy conservation. In quantum transitions, anything not specifically forbidden *will* happen, with a specific transition probability.

We can now answer the question at the end of Chap. 5.

Q. How does an electron emit light by jumping from one state to another?

A. The electron was initially in an excited state corresponding to some Bohr orbit. The quantized electromagnetic field fluctuates about the value zero in the vacuum. These quantum fluctuations buffet the electron continually, and cause the electron to decay to the ground state by emitting a photon, with a certain probability amplitude.

10.3. Virtual photons

If two electrons are near each other, then the photon emitted by one can be absorbed by the other almost immediately. The photon exists only briefly in a "virtual" state.

Fig. 10.5 A real photon has zero mass, and can propagate indefinitely. A virtual photon has non-zero mass, and damps out as it propagates.

If the virtual photon lasts for a time Δt, its energy has an uncertainty $\Delta E \approx \hbar/\Delta t$. That is to say, in emitting or absorbing a virtual photon, energy is conserved only to the extent ΔE. In Lorentz-invariant language, we can rephrase this by saying that energy is conserved, but the virtual photon has a mass different from zero.

That is, a virtual photon "goes off the mass shell".

As illustrated in Fig. 10.5, a real photon, whose mass is zero, can propagate indefinitely. A virtual photon with non-zero mass, on the other hand, damps out as it propagates, with a mean life inversely proportional to the virtual mass.

An electron trying to absorb a virtual photon must be close enough to catch it before it dies. This creates an interaction between the emitter and the absorber. This is the origin of the electromagnetic interaction between electrons.

Creation and Annihilation

11.1. The quantum field

An electron can emit a photon, which did not exist beforehand, and was created in the emission process. Similarly, a photon was annihilated when absorbed by an electron, and no trace remains. Such behaviors fall outside of the Schrödinger equation, which conserves particle number. It has to be described through quantum field theory.

A quantum field is the quantized version of a classical field, and consists of operators attached to each point of space-time. The electromagnetic field is described by the vector potential $A^j(x)$, where x denotes a space-time coordinate, and $j = 1, 2, 3$ labels spatial vector components. These were real numbers in classical theory, and become hermitian operators upon quantization.

A state of the electromagnetic field can be specified by enumerating all the photons present, and a list of states looks something like this:

$$\begin{aligned}
\text{Vacuum state:} \quad &|0\rangle \\
\text{1-photon states:} \quad &|\gamma_1\rangle \\
\text{2-photon states:} \quad &|\gamma_1, \gamma_2\rangle \\
\text{3-photon states:} \quad &|\gamma_1, \gamma_2, \gamma_3\rangle \\
&\vdots \qquad\qquad \vdots
\end{aligned}$$

where γ summarizes the momentum and polarization of a photon. Each line in the table corresponds to a subspace with a fixed number of photons. The unique vacuum state has no photon.

The role of the field operator is to connect adjacent subspaces, by annihilating or creating one photon. Specifically, $A^j(x)$ annihilates or creates a photon at the space-time point x, and the vector index j denotes the direction of linear polarization of the photon. Thus, the field operator contains two terms: one for annihilation, and the other for creation.

Hiding j and x for simplicity, we can write

$$A = A^{(-)} + A^{(+)}\,,$$

where the two terms are hermitian conjugates of each other, and

$A^{(-)}$ annihilates a photon at a space-time point,

$A^{(+)}$ creates a photon at a space-time point.

When we take Fourier transforms, space-time goes into 4-momentum. The Fourier transform of A, denoted by $\tilde{A}$, has the decomposition

$$\tilde{A} = a + a^\dagger,$$

where $\dagger$ denotes hermitian conjugate, and

a annihilates a photon of given momentum,

$a^\dagger$ creates a photon of given momentum.

By applying the creation operators $a^\dagger$ repeatedly on the vacuum state, we can create a state with any number of photons:

$$|\text{Photons}\rangle = (a^\dagger a^\dagger a^\dagger \cdots)\,|0\rangle\,.$$

We can annihilate them back into the vacuum:

$$|0\rangle = (aaa\cdots)\,|\text{Photons}\rangle\,.$$

The states above are defined up to a multiplicative constant. If there are no photons to annihilate, the action of the annihilation operator a gives zero.

11.2. Particle and antiparticle

According to relativity, every particle comes with an antiparticle.

In relativistic mechanics, the energy of a particle E is related to its momentum p through $E^2 = p^2 + m^2$, where m is the rest mass.[1] Since the square root can be either positive or negative, the energy has two signs:

$$E = \pm\sqrt{p^2 + m^2}.$$

In quantum theory, energy corresponds to the frequency of matter waves (times Planck's constant), and both signs have to be taken into account. For a given magnitude of frequency $\omega = |E|/\hbar$, the wave function has two branches with the following time dependences:

$$\exp(-i\omega t) \qquad \text{(Positive frequency part)},$$
$$\exp(i\omega t) \qquad \text{(Negative frequency part)}.$$

By convention, the former refers to a particle, and the latter an antiparticle.

The antiparticle needs not be distinct from the particle. The photon is its own antiparticle. For other particles such as the electron or proton however, the particle and antiparticle are different.

When a particle meets its antiparticle, both disappear in a puff of energy (or, as they say, gamma rays). We can reverse the process: gamma rays can create a particle–antiparticle pair from the vacuum.

In relativistic theory, particles and antiparticles can be created or annihilated. We are necessarily dealing with varying numbers of particles, and this calls for quantum field theory.

When particle and antiparticle are different, the quantum field operator Ψ is different from its hermitian conjugate $\Psi^\dagger$:

Ψ annihilates a particle, or creates an antiparticle, at a space-time point,

$\Psi^\dagger$ creates a particle, or annihilates an antiparticle, at a space-time point.

---

[1] From now on, the term mass shall always mean rest mass. Accordingly, we drop the subscript in m_0. We also use the velocity of light as the unit of velocity. Thus, $c = 1$, v/c becomes v, and mc^2 becomes m.

The Fourier transforms have the forms

$$\tilde{\Psi} = a + b^\dagger,$$
$$\tilde{\Psi}^\dagger = a^\dagger + b,$$

where

> a annihilates a particle of given momentum,
>
> $a^\dagger$ creates a particle of given momentum,
>
> b annihilates an antiparticle of given momentum,
>
> $b^\dagger$ creates an antiparticle of given momentum.

11.3. The Dirac equation

The Schrödinger equation is inherently non-relativistic, since space and time play distinct roles. A relativistic equation should be covariant under the Lorentz transformation, which mixes space and time.

Dirac tackled the problem in 1928 by postulating an equation of the form

$$\left(i\gamma^\mu \partial_\mu - \frac{mc}{\hbar} \right) \psi = 0,$$

where γ^μ are objects to be determined. In order for this equation to conform to the relativistic relation between energy and momentum, he found that γ^μ should be 4×4 matrices with specific properties. Thus, the wave function ψ must have 4 components:

$$\psi = \begin{pmatrix} \psi_1 \\ \psi_2 \\ \psi_3 \\ \psi_4 \end{pmatrix}.$$

The components have the following meaning:

- ψ_1 and ψ_2 represent states of a spin $1/2$ particle.
- ψ_3 and ψ_4 represent states of negative energy, as required by relativity.

Wolfgang Pauli (1900–1958)

Fig. 11.1 The Pauli exclusion principle forbids two electrons to occupy the same state, so they can stack up in atomic levels to give rise to the periodic table, and in negative-energy states in the vacuum to form the Dirac sea.

This shows that spin is an intrinsic property that cannot be "tacked on" as an afterthought, as done in non-relativistic theories.

The existence of the negative-energy states poses a potential disaster.

11.4. The Dirac sea

The negative-energy spectrum has no bottom. This seems at first glance to be disastrous, for particles can keep dropping in energy, disappearing down the bottomless pit. Dirac saved the situation by making the bold assumption that the vacuum is the state in which all negative-energy states are filled.

What makes the vacuum stable is the Pauli exclusion principle, which states that no two electrons can occupy the same state. This principle was originally proposed to enable the successive filling of

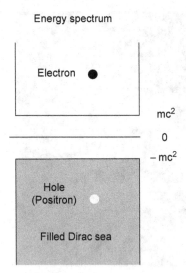

Fig. 11.2 The vacuum state is the completely filled Dirac sea. A hole in the sea is a positron, antiparticle to the electron.

electron orbits in atoms, so as to construct the periodic table. Dirac seized upon this to stabilize the vacuum.

The completely filled negative-energy states is called the Dirac sea, which is invisible by definition. However, any deviations from it will be observable as an excitation from the vacuum. In particular, an absence of an electron in the sea — a hole — would be seen as a particle of the same mass as the electron but with opposite charge — the antiparticle. An electron jumping into the hole to refill it will be seen as the annihilation of an electron–positron pair. This is illustrated in Fig. 11.2.

The electron's antiparticle, called the positron, was discovered by Carl Anderson in 1932. The positronium, an atom made up of electron and positron (instead of the proton), was discovered by Martin Deutsch in 1951.

11.5. Reversing time

We saw that particle and antiparticle are associated with opposite signs of the frequency. Thus, they go into each other under time

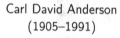

Carl David Anderson   Martin Deutsch
(1905–1991)   (1917–2002)

Fig. 11.3 In 1932, Carl Anderson discovered the positron — the hole in the Dirac sea. In 1951, Martin Deutsch discovered positronium, bound state of electron and positron with a lifetime of a nanosecond.

reversal. This underlies Feynman's picture that an antiparticle is a particle moving backwards in time.

Feynman recalled a telephone conversation with his Ph.D. thesis supervisor, John Wheeler in 1947:[2]

> I received a telephone call one day at the graduate college at Princeton from Professor Wheeler, in which he said,
> "Feynman, I know why all electrons have the same charge of the same mass."
> "Why?"
> "Because they are all the same electron!"
> And then he explained on the telephone,
> "(If an electron's world line) reversed itself, and is coming back from the future, we have the wrong sign to the proper time, and that is equivalent to changing the sign of the charge, and therefore that part of a path world act like a positron."

---

[2]R. P. Feynman, "The development of the space-time view of quantum electrodynamics", in *Les Prix Nobel 1965*, (Imprimerie Royale P. A. Norsredt and Soner, Stockholm, 1966), pp. 172–191.

John A. Wheeler (1911–2008)

Fig. 11.4   John A. Wheeler put the idea into his student Feynman's head, that a positron is an electron moving backwards in time.

"But, Professor", I said, "there aren't as many positrons as electrons."

"Well, maybe they are hidden in the protons or something", he said.

I did not take the idea that all the electrons were the same one from him as seriously as I took the observation that positrons could simply be represented as electrons going from the future to the past in a back section of their world lines. That, I stole!

11.6. Feynman diagram

Using the idea that the positron is an electron moving backwards in time, Feynman introduces space-time diagrams in his "theory of positrons".[3] The original pictures and captions in his paper are reproduced in Fig. 11.5. A line with an arrow denotes a world line. An electron travels along the direction of time, while a positron travels backward in time. Interactions take place in the circled regions.

[3]R. P. Feynman, *Phys. Rev.* **76**, 749 (1949).

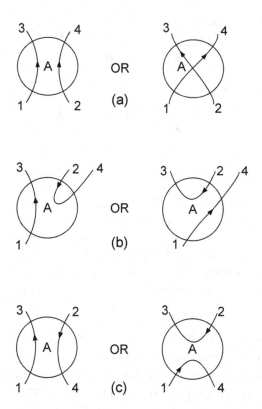

Fig. 11.5 These space-time diagrams, with time flowing upward, illustrate the idea that an antiparticle is a particle traveling in a reversed direction of time. The following are original captions by Feynman: (a) Electrons at 1 and 2 are scattered to 3,4. (b) Starting with an electron at 1 a single pair is formed, positron at 2, electrons at 3,4. (c) A pair at 1,4 is found at 3,2.

The different space-time diagrams are really one diagram with different choices of initial and final states. They have been distilled into the Feynman diagrams we use today, as shown in Fig. 11.6. There is no longer a time direction, and they are just graphical shorthands for scattering amplitudes.

A dot, called a vertex, marks the basic event: emission of absorption of a photon by an electron. A directed line represents a particle: it is an electron if the momentum is directed along the

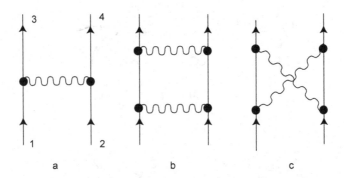

Fig. 11.6 Feynman diagrams, distilled from space-time diagrams, are shorthands for mathematical expressions of scattering amplitudes. Here, the amplitudes for ee, e$\bar{\text{e}}$, $\bar{\text{e}}\bar{\text{e}}$ scattering are all represented by the same diagram. Which process is being described depends on the choice of initial channel. (e = electron, $\bar{\text{e}}$ = positron). The different diagrams represent contributions from different virtual processes involving the exchange of photons.

arrow, a positron if against the arrow. A wavy line represents a photon, which is not directed, because the photon is its own antiparticle.

Figure 11.6(a) represents the lowest, second-order diagram, in which the interaction happens twice. The others [Figs. 11.6(b) and (c)] are fourth-order. The electromagnetic interaction is relatively weak, and higher order diagrams make increasingly small contributions. In the lowest order process, the two particles scatter via exchange of one virtual photon. The others represent contribution from higher order iterations.

The same Feynman diagram represents different processes, depending on the choice of the initial channel. Once that is done, the final channel is automatically determined. For example (with e = electron, $\bar{\text{e}}$ = positron):

$$12 \rightarrow 34 \qquad \text{ee scattering}$$
$$13 \rightarrow 24 \qquad \text{e}\bar{\text{e}} \text{ scattering}$$
$$14 \rightarrow 23 \qquad \text{e}\bar{\text{e}} \text{ scattering.}$$

The time-reversed reaction, such as $34 \rightarrow 12$, corresponds to the reaction with particle and antiparticle interchanged.

Each diagram corresponds to a specific matrix element, which can be written down using "Feynman rules". The exact transition matrix is the sum of all possible Feynman diagrams.

The Feynman diagram has worked such magic that pre-Feynman calculations filling pages have been reduced to one line. More than that, a set of Feynman rules enables one to calculate any scattering amplitude, and therefore defines a quantum field theory.

11.7. The fine-structure constant

The charge of the electron acts as a coupling constant in the electron–photon interaction. A vertex in a Feynman diagram is associated with a "bare" coupling constant e_0, which gets "renormalized" to the physical charge e through interaction effects (Chap. 12). The observed value of e is contained in a dimensionless combination called the *fine-structure constant*

$$\alpha = \frac{e^2}{\hbar c} \approx \frac{1}{137.040}.$$

This name comes from the fact that it was first measured in the splitting of atomic spectral lines called "fine structure".

The smallness of α makes it possible to calculate scattering amplitudes in successive approximation, by expanding in powers of α. The procedure is called *perturbation theory*, and Feynman diagrams are ideal for that.

It is striking that α^{-1} is so close to the prime number 137. We have no clue why this is so, but it never ceases to inspire awe and mystical speculation.[4]

---

[4]The renowned British astronomer Sir Arthur Stanley Eddington (1882–1944), allegedly one of only two people in the whole world who understood Einstein's theory of general relativity, discovered that the number of degrees of freedom of his universe was precisely 137.

The Dynamical Vacuum

12.1. QED

QED (quantum electrodynamics) is the relativistic quantum field theory of interacting electrons and photons. It consists of Dirac's electron coupled to Maxwell's gauge field, in the framework of quantum field theory.

Because of interactions, the vacuum becomes a cauldron of fluctuating fields. Not only are the electric and magnetic fields fluctuating, but the Dirac sea also fluctuates, with spontaneous creation and annihilation of virtual electron–positron pairs. These vacuum fluctuations have observable effects that can be calculated in QED with the help of Feynman diagrams.

The electromagnetic coupling is measured by the fine-structure constant $1/137$, which can be treated as a small parameter. It gives rise to "radiative corrections" to properties of the free electron and the free photon. There are three basic processes that we shall describe separately:

- Vertex correction,
- Electron self-energy,
- Vacuum polarization.

12.2. Interaction vertex

The diagrams in Fig. 12.1 show how the unperturbed "bare" vertex gets modified by the lowest order radiative correction. While the bare

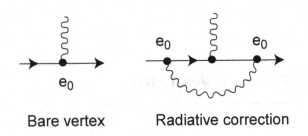

Bare vertex   Radiative correction

Fig. 12.1   The bare vertex represents a basic element in Feynman diagrams, at which an electron emits or absorbs a photon. The radiative correction gives it structure, and contributes to the "anomalous magnetic moment" of the electron.

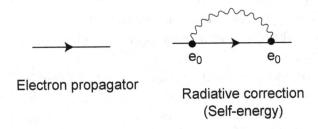

Electron propagator   Radiative correction (Self-energy)

Fig. 12.2   Radiative corrections to the electron propagator describe self-interaction that contributes to mass renormalization.

interaction occurs at one point, the correction smears it out over a region.

There are higher order diagrams that will improve the accuracy of the calculation. When all possible Feynman diagrams are included, the electron is seen to emit a photon from within a "blob", which contains the electron's structure endowed by QED. Attributes of the structure include the "anomalous magnetic moment" that we shall describe later.

12.3. Self-energy

Figure 12.2 shows the "bare" propagator of the electron, which is represented by a directed line, and is a building block of Feynman

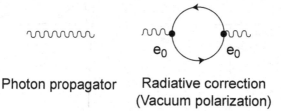

Photon propagator   Radiative correction
(Vacuum polarization)

Fig. 12.3 Vacuum polarization: a propagating photon can momentarily materialize into a virtual electron–positron pair, thus producing charge separation in the vacuum.

diagrams. Physically it describes the probability amplitude that an electron created at point A can reach point B.

The radiative correction to the bare propagator involves the emission and absorption of a virtual photon by an electron. This corresponds to self-interaction of the electron. This and higher order corrections lead to a "full" propagator that describes a "dressed" electron.

The bare propagator contains the bare mass, a parameter in the QED Hamiltonian. The radiation corrections yield a "self-energy" corresponding to a mass correction:

Physical mass = (Bare mass) + (Self-mass).

This formula expresses what is known as *mass renormalization*.

12.4. Vacuum polarization

Photon self-energy graphs are shown in Fig. 12.3. In the lowest order correction to the unperturbed propagator, the photon creates a virtual pair from the vacuum, which annihilates, re-emitting the photon. The momentary charge separation endows the vacuum with a distribution of induced electric dipole moments, and the process is called "vacuum polarization".

The photon mass cannot change, because it is kept at zero by gauge invariance. The chief effect of vacuum polarization is to alter the electron's charge distribution as seen by an external probe. It leads to charge renormalization, as discussed below.

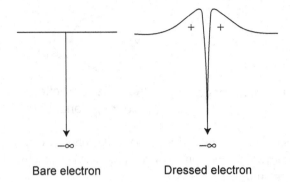

Bare electron   Dressed electron

Fig. 12.4 Vacuum polarization "dresses" the electron with induced charges. The screening of the bare charge results in an effective charge dependent on distance from the center, leading to the designation "running coupling constant". The total charge seen at infinity, the renormalized (physical) charge, has a magnitude smaller than the bare charge.

12.5. The dressed electron

Two electrons exchanging a virtual photon are probing each other's charge distribution. Vacuum polarization by the photon "dresses" the electron being probed with an induced dipole distribution, and screens the bare charge. This is depicted in Fig. 12.4.

As we go away from the center of the charge distribution, the total charge seen by a probe becomes smaller. The effective charge of the electron varies with distance from the center, and is called a "running coupling constant" for this reason. (More in Chap. 21.)

The screening length is microscopic, of order

$$\frac{\hbar}{mc} \approx 4 \times 10^{-11} \text{ cm}.$$

Outside of this distance, the effective charge rapidly converges to a limiting value, the physical charge:

Physical charge = (Bare charge) × (Sceening factor).

This is called *charge renormalization*. Note that it is multiplicative, whereas mass renormalization is additive.

John C. Ward (1924–2000)

Fig. 12.5 Charge renormalization is due solely to vacuum polarization by the photon, and therefore universal.

In 1950, John Ward showed that charge renormalization is universal, in that the screening factor depends only on vacuum polarization produced by the photon, and is therefore the same for all charged particles. The mathematical relation implying this universality is known as *Ward's identity*.

12.6. The ultraviolet catastrophe

In calculating scattering amplitudes using Feynman diagrams, one finds that some integrals involved are divergent, due to contributions from high-frequency modes. By cutting off the integrations at some high frequency, one can calculate the amplitudes as functions of the cutoff. The trouble is that they become infinite when the cutoff approaches infinity. This is called the *ultraviolet catastrophe*.

Eventually the debacle was circumvented through *renormalization*.

In 1949, Freeman Dyson proved that all divergent integrals that we could ever encounter can be absorbed into mass and charge

Freeman J. Dyson (1923–)

Fig. 12.6 All divergences in QED can be absorbed into mass and charge renormalization.

renormalization. That is, the cutoff appears only within the self-mass, or the charge-screening factor. We can thus take the values of the physical mass and charge from experiments, and forget about the cutoff.

A theory like QED, in which divergences can be absorbed into physically measurable quantities such as mass and charge, is said to be *renormalizable*. It means that the theory is self-similar: when the cutoff is changed, only observable parameters like charge and mass change, while the forms of scattering amplitudes remain the same.

People had imagined that they could hold the renormalized mass and charge fixed, and "send the cutoff to infinity", thereby restoring the theory to an "unblemished" state. In reality, this can be done only in some theories (so-called asymptotically free theories), and that does not include QED.

It took a few decades before we could understand the physical meaning of renormalization, that it corresponds to a change of resolving power in viewing the system. This realization does not change the way we deal with divergent Feynman diagrams; we still use essentially the same technique codified by Dyson and others more than fifty years ago. However, it helps us to understand the true meaning of theory. We shall discuss this topic at the end of the book (Chaps. 21 and 22).

12.7. Reality of vacuum fluctuations

We now recount some historic events in the development of QED.

The motivation to study QED came from experiments that measure observable effects of vacuum fluctuations.

In the hydrogen atom, an electron in an S state (with circular orbit) and P state (with figure-8 orbit) have the same energy in Dirac's electron theory, in the absence of vacuum fluctuations. Fluctuations of the quantized electromagnetic field in the vacuum cause a splitting of their energies, because the orbits are being distorted in slightly different manners. The splitting amounts to one part in a million, as indicated in Fig. 12.7. This was measured by Willis Lamb in 1947, and is named the "Lamb shift".

Another observable effect is the "anomalous magnetic moment" of the electron.

Particles with spin S have a magnetic moment[1]

$$\mu = gS,$$

and the factor g is called the gyromagnetic ratio. For an electron, with $S = 1/2$, the Dirac equation predicts $g = 2$, and this result can be tested to high accuracy. As illustrated schematically in Fig. 12.7, when an electron is placed in a uniform magnetic field, it moves in a circular orbit about the direction of the field, and its spin precesses about the same direction. If $g = 2$, the spin precession is precisely synchronized with orbital motion. To test this value, all one needs to

---

[1] The spin is measured in units of $\hbar$, while the magnetic moment is given in units of the Bohr magneton $e\hbar/2m$.

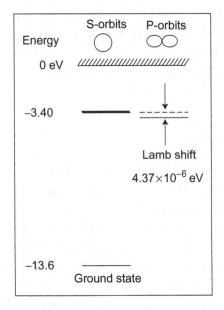

Willis E. Lamb
(1913–2008)

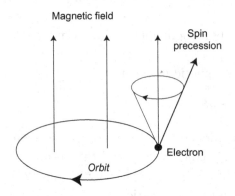

Polykarp Kusch
(1911–1993)

Fig. 12.7 Lamb shift (upper panel) and anomalous magnetic moment of electron (lower panel). The Lamb shift is due mainly to the difference in self-energy of an electron in the 2S and 2P states of hydrogen, and amounts to one part in a milion. The drawing in lower panel shows an electron in a circular orbit in a uniform magnetic field, and its spin precesses about the magnetic field. According to Dirac theory, these two periodic motions should be prefectly synchronized. Due to the vertex correction however, the spin precession slips behind the orbital by one part in a thousand.

Fig. 12.8 Richard Feynman (seated, with pen in hand) explains a point at the Shelter Island conference (1947). From left to right, standing: Willis E. Lamb, K. K. Darrow, Victor F. Weisskopf, George E. Uhlenbeck, Robert E. Marshak, Julian S. Schwinger, David Bohm. From left to right, seated: J. Robert Oppenheimer (holding pipe), Abraham Pais, Richard P. Feynman, Herman Feshbach. (Image credit: National Academy of Sciences.)

do is to observe the electron for a long time, and detect any slippage between the two periodic motions.

Vacuum fluctuations, in the form of vertex correction, will contribute an "anomalous magnetic moment" and make g deviate from 2. The theoretical result, one of the early predictions of QED, is called the Schwinger correction:

$$g - 2 = \frac{\alpha}{\pi}.$$

This was experimentally verified by Polykarp Kusch in 1947, at about the same as the Lamb shift.

Hans A. Bethe   Julian S. Schwinger   Victor F. Weisskopf
(1906–2005)   (1918–1994)   (1908–2002)

Fig. 12.9 Bethe, Schwinger, Feynman, and Weisskopf (with student J. Bruce French) calculated the Lamb shift independently during 1947–1948, using renormalization to circumvent the "ultraviolet catastrophe".

12.8. When physicists were heroes

Calculation of the Lamb shift proved to be more challenging than the Schwinger correction, because here one faces the ultraviolet catastrophe. In June 1947, a conference was sponsored by the U.S. National Science Academy at Shelter Island, NY, to discuss this and related problems. It was attended by 24 physicists, many freshly returned from Los Alamos, where they had worked on the atomic bomb that ended World War II.

The idea of renormalization emerged at the conference. Using this idea, Hans Bethe made a rough calculation of the Lamb shift, reportedly on the train back to Cornell University from the conference. Shortly thereafter, the Lamb shift was successfully calculated by three people independently: Julian Schwinger, using elegant operator techniques he had developed; Richard Feynman, using his space-time approach; and, most arduously, Victor Weisskopf, assisted by his graduate student, J. Bruce French, using "old-fashioned" techniques.

The seminal Shelter Island Conference took place at a time when physicists were heroes in America. Here's how it was:[2]

> The attendees were treated like celebrities when they arrived at Greenport, Long Island, where they stopped before heading on to Shelter Island. John C. White, president of the Greenport Chamber of Commerce and a Marine in the Pacific in WWII, arranged and paid for a dinner for the visiting scientists out of gratitude for the war work done by the physicists who developed the atomic bomb. One conferee recalled that during their trip to Greenport, the group was given a series of motorcycle police escorts and their bus was allowed to run through red lights.

12.9. The enduring QED

Renormalization has skirted the ultraviolet catastrophe through a mathematical recipe. Unexpectedly, it has made QED the most precise theory in all of physics.

In 2006, a group at Harvard University led by Gerald Gabrielse made the most accurate measurement of the electron's anomalous magnetic moment to date, by suspending a single electron for months in a trap. They achieved the incredible precision of one part in a trillion:

$$\frac{1}{2}g_{\text{expt}} = 1.00115965218085(76) \, .$$

Every significant figure in this result agrees with theoretical calculations in QED, a lifelong project of Toichiro Kinoshita:[3]

---

[2] From http://www7.nationalacademies.org/archives/shelterisland.html.
[3] *Physics Today*, August (2006), p. 15; G. Gabrielse and T. Kinoshita *et al.*, *Phys. Rev. Lett.* **97**, 030802 (2006).

Toichiro Kinoshita (1925–)

Fig. 12.10 Toichiro Kinoshita, whose life work culminated in the calculation of the electron anomalous moment to 8th order. The result agrees with experiment to a precision of one part in a trillion.

Here are the historical theoretical results, with year of publication and number of Feynman diagrams computed:[4]

$$
\begin{aligned}
\tfrac{1}{2} g_{\text{theory}} = \quad & 1 & \text{(a) 1928 (Dirac equation)} \\
& + (\alpha/2\pi) & \text{(b) 1949 (1 diagram)} \\
& -0.32848\,(\alpha/\pi)^2 & \text{(c) 1958 (18 diagrams)} \\
& + (1.195 \pm 0.026)\,(\alpha/\pi)^3 & \text{(d) 1974 (72 diagrams)} \\
& - (1.7283\,(35))\,(\alpha/\pi)^4 + (\text{Non-QED}) & \text{(e) 2006 (891 diagrams).}
\end{aligned}
$$

The non-QED contribution arises from the weak and strong interactions. This result is so precise that, through comparison with experiment, one can obtain the most accurate determination of the fine structure constant so far:

$$
\frac{1}{\alpha} = 137.035999710(96)\,.
$$

[4](a) From Dirac equation; (b) J. Schwinger, *Phys. Rev.* **75**, 651 (1949); (c) C. M. Summerfield, *Ann. Phys.* (*NY*) **5**, 26 (1958); (d) P. Cvitanovic and T. Kinoshita, *Phys. Rev. D* **10**, 4007 (1974); (e) T. Kinoshita and M. Nio, *Phys. Rev. D* **73**, 013003 (2006).

Dyson, in a letter of congratulation to Gabrielse,[5] wrote:

We thought of QED as a jerry-built structure. We didn't expect
it to last more than 10 years before a more solidly built theory
replaced it. But the ramshackle structure still stands. The revealing discrepancies we hoped for have not yet appeared. I'm
amazed at how precisely Nature dances to the tune we scribbled
so carelessly 57 years ago, and at how the experimenters and
theorists can measure and calculate her dance to a part in a
trillion.

The electron has been treated in QED as a point charge "dressed"
by interactions. Experimental discrepancies from the predictions of
QED will indicate the existence of intrinsic structures not taken into
account so far. The agreement with QED so far sets an experimental
upper limit of 10^{-16} cm for the intrinsic radius.

Kinoshita is pushing on to the 10th-order doggedly, with over ten
thousand Feynman diagrams to calculate. All await the day when a
discrepancy with experiment is found, and a new ball game begins.

[5] *Physics Today*, August (2006), p. 17.

13

Elementary Particles

13.1. Beginnings

The first elementary particles were the electron, discovered by J. J. Thomson in 1897, and the proton, discovered by Ernest Rutherford in 1911.

J. J. Thomson measured the charge-to-mass ratio of the electron in the cathode-ray tube shown in Fig. 13.2. The experiment revealed the granular nature of electricity for the first time. The model of the atom then consisted of electrons embedded in a uniform background of positive charge, to make the system electrically neutral.

In 1911, Rutherford demonstrated that the positive charge in the atom was concentrated in a small nucleus at the center. The nucleus of hydrogen was identified as the proton, with a mass approximately 2000 times that of the electron. Nearly thirty years passed when, in 1932, James Chadwick discovered the neutron — the neutral component of the nucleus of about the same mass as the proton.

Beta radioactivity was discovered around 1900, in which a nucleus decays into another one by emitting an electron. The decaying state had a very long lifetime, indicating that the interaction was very weak, and was consequently called the *weak interaction.*

If the final state in beta decay consisted solely of nucleus and electron, their energies should have fixed values, as dictated by energy and momentum conservation. The observed energy of the electron, however, exhibits a continuous spectrum of values. This had caused

J. J. Thomson   Ernest Rutherford
(1856–1940)   (1871–1937)

Fig. 13.1 The pioneers: Thomson discovered the electron in 1897; Rutherford discovered the proton in 1911.

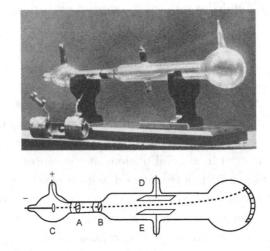

Fig. 13.2 J. J. Thomson's cathode-ray tube. An electron beam produced at C was collimated by slits at A and B. The electron's charge-to-mass ratio can be determined by deflecting the beam with electric and magnetic fields between D and E.

great puzzlement. Niels Bohr even entertained the idea that energy was not conserved.

Wolfgang Pauli made the bold suggestion in 1930, that an extra particle must have accompanied the electron in beta decay. This

Fig. 13.3 The neutrino was postulated by Wolfgang Pauli in 1930, and named and utilized in a theory by Enrico Fermi in 1933. Eluding detection for over two decades, it was found at last in 1956, by Frederick Reines (1918–1998) and Clyde Cowan (1919–1974), shown here in the thick of their experiment.

particle would have zero charge and mass, and thus escape detection. Enrico Fermi used this particle in his 1933 theory of beta decay, and dubbed it the *neutrino* — the little neutron. Fermi's theory quickly became accepted, but the neutrino continued to elude detection until 1956, when it was finally caught by Clyde Cowan and Frederick Reines, pictured working together in Fig. 13.3.

Two particles showed up in cosmic rays from outer space: the μ- and π-mesons, or muon and pion, respectively.

The π meson was proposed by Yukawa in 1934 as the mediator of the strong interaction, modeled after the photon that mediates the electromagnetic interaction. From the size of the nucleus, Yukawa estimated the mass to be about 100 MeV. It was discovered by Cecil Powell in 1947, by reading particle tracks produced by cosmic rays in photographic emulsions. Its actual mass is 140 MeV, close to Yukawa's prediction.

James Chadwick　　Hideki Yukawa　　Cecil Powell
(1891–1974)　　　(1907–1981)　　　(1903–1969)

Fig. 13.4　Chadwick discovered the neutron in 1932. Yukawa predicted the π meson in 1934, and Powell observed it in cosmic rays in 1947.

Fig. 13.5　"Who ordered the μ-meson?" — Isidor Isaac Rabi (1898–1988).

We now know that the true intermediary for the strong interaction is not the pion, but Yang–Mills photons (Chap. 17), but the pion plays a singularly intriguing role that is not completely understood (Chap. 20).

The muon was observed in 1936 by Carl Anderson, discoverer of the positron. It was an enigma from the beginning. ("Who ordered it?" asked I. I. Rabi, father of the molecular beam.) Since then, it has acquired siblings, and now the whole family has become the enigma (Chap. 17).

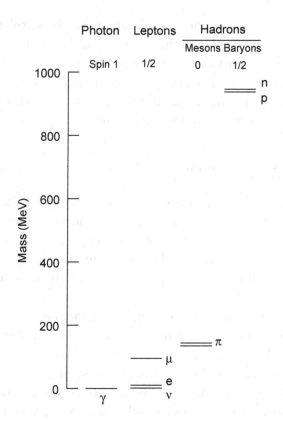

Fig. 13.6 The elementary particle spectrum as of 1947, the triumphant year of QED.

In Fig. 13.6 we display the known spectrum of elementary particles as of 1947, the triumphant year of QED. They were classified according to weight — heavy, medium, light — as baryon, meson, lepton. Current usage associates leptons with the weak interaction, and introduces the name "hadron" to denote strongly interacting particles. The muon is akin to the electron rather than the pion, and is now classified as a lepton.

These particles build a neat world in an energy range in which the nucleus appears to be passive and inert. The active players are atoms, made of electrons bound to the nucleus by the electrostatic Coulomb attraction. The same Coulomb interaction working between different atoms leads to chemical bonding and chemical reactions. These phenomena, together with emission and absorption of photons, account for a large part of our everyday world.

Beyond this world, a few steps away in terms of energy, lies a strange lava dome waiting to erupt. Before we peer into the cauldron, let us review some basic properties of elementary particles.

13.2. Bosons and fermions

Particles of the same type are identical, indistinguishable in principle. The indistinguishability is a quantum mechanical property foreign to classical physics. Two particles are identical if the Hamiltonian is invariant under an interchange of their coordinates. This is true both in classical as well as quantum mechanics, but the common thread ends here.

In classical mechanics, the coordinate is a number with which you tag a particle. You can always distinguish one particle from another by looking at their tags.

In quantum mechanics, by contrast, the coordinate is an operator. A two-particle system is described by a wave function dependent on two coordinates, but you cannot tell which coordinate belongs to which particle. The reason is as follows.

The wave function of a stationary state is an eigenfunction of the Hamiltonian. Under an interchange of identical particles, the Hamiltonian remains invariant, and therefore the wave function must remain an eigenfunction. That is, interchanging particles take us from an eigenstate to an eigenstate, which may or may not be the same as the original one.

If there are degeneracies, that is, if a group of states have the same energy, then the degenerate states could mix under particle exchange. As far as we know, however, this does not happen in physical systems, and particle interchange leaves the state unchanged. That means the wave function can change at most by a multiplicative factor.

Since performing the exchange operation twice brings you back to the initial situation, the wave function either remains unchanged or changes sign.

Satyendra Nath Bose   Enrico Fermi
(1894–1974)   (1901–1954)

Fig. 13.7 Discoverers of Bose statistics and Fermi statistics. Particles obeying Bose statistics are called bosons, and they like to congregate in the same state. Fermions, which obey Fermi statistics, are forbidden to occupy the same state.

In a system of many identical particles, the wave function is either symmetric under an interchange of a pair of particle coordinates, or antisymmetric. This characteristic is called the "statistics" of the particles. The symmetric case corresponds to "Bose statistics", and the antisymmetric case corresponds to "Fermi statistics", after their discoverers S. N. Bose and Enrico Fermi, respectively. Particles with these properties are called "bosons" and "fermions", respectively.

Boson like to congregate in the same single-particle state, whereas two fermions cannot be in the same single-particle state. The latter property is known as the Pauli exclusion principle. The state of lowest energy for a group of bosons will have all the particles in the lowest level. For fermions, however, they stack up on the levels one by one. This is illustrated in Fig. 13.8.

The effective "Bose attraction" and "Fermi repulsion" have important physical consequences. The former rewards photons for having the same frequency, and underlies the principle of the laser. The latter stabilizes the electron vacuum state — the Dirac sea.

For matter in bulk, the Bose attraction gives rise to Bose–Einstein condensation, which leads to superfluidity and superconductivity. In

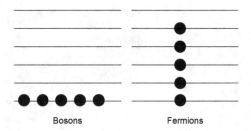

Bosons   Fermions

Fig. 13.8 Bosons prefer to be in the same state, whereas fermions are forbidden to do that.

metals, the analog of the Dirac sea (the top of which is called the Fermi surface) underlies the distinctive behaviors of conductors, insulators, and semiconductors.

13.3. Spin and statistics

Experimentally, particles of integer spin $(0, 1, 2, \ldots)$ are bosons, while those of half-integer spin $(1/2, 3/2, \ldots)$ are fermions. This rule not only holds for elementary particles, but for composite states as well. Thus, a group of bosons bind into a boson. For fermions, an odd-numbered group binds into a fermion, while an even-numbered group binds into a boson.

For example, the nucleus of N (nitrogen) is of mass 14 and charge 7, in proton units. It is composed of 7 protons and 7 neutrons, and is therefore a boson. Before the discovery of the neutron, however, N was thought to be composed of 14 protons and 7 electrons. This made it a fermion, because of the odd number of fermions. The statistics leads to essential physical differences, for it determines the symmetry character of the molecule N_2. The molecular wave function must be symmetric under interchange of the two atoms in the case of Bose statistics, and antisymmetric in the case of Fermi statistics. Through observation of the rotational spectrum of N_2, early molecular spectroscopy revealed that the wave function of N_2 is symmetric, and therefore it should be a boson. In this manner, the existence of the neutron was anticipated before its actual discovery.

The Dirac sea owes its stability to the fact that the spin-1/2 electron is a fermion, and thus obeys the Pauli exclusion principle.

The spin–statistics connection follows from Lorentz invariance in a "local" quantum field theory. As Feynman[1] pointed out however, a more fundamental understanding lies in the mathematical relation between rotation and position exchange.

13.4. Interactions

In the particle spectrum of Fig. 13.6, the photon stands alone as the massless vector boson. The other particles were originally classified by weight:

- Baryon (heavy): proton and neutron,
- Meson (intermediate): muon and pion,
- Lepton (light): electron and neutrino.

The usage has changed over the years, however. Lepton now denotes a particle with weak interaction, and the muon moves into that category.

We know of four types of interactions, which are listed below in decreasing order of strength:

Interaction	Strength	Range	Charge neutrality?
Strong	10	Finite	Yes
Electromagnetic	10^{-2}	∞	Yes
Weak	10^{-5}	Finite	Yes
Gravitational	10^{-36}	∞	No

We note the following features:

- Particles having the strong interaction are called "hadrons", while the "lepton" is a particle without strong interaction, with lepton

[1]R. P. Feynman and S. Weinberg, *Elementary Particles and the Law of Physics; The 1986 Dirac Memorial Lectures* (Cambridge University Press, New York, 1987).

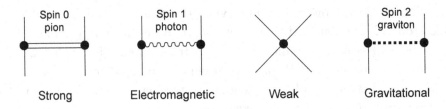

Fig. 13.9 In early views, the weak interaction was a contact interaction, and the strong interaction was mediated by the pion. In the modern view, all but gravity are mediated, by spin-1 vector bosons. The graviton is yet to be understood.

number conservation.[2] The muon, now a lepton, is some 200 times heavier than the electrons. Leptons to be discovered later are much heavier than the proton.

- The name "baryon" now refers to fermions which obey baryon number conservation, like nucleons. "Meson" now refers to particles whose number is not conserved, like pions.

- The strong interaction is very strong, but has a range of nuclear size, or 10^{-13} cm. The weak interaction is not only a thousand times weaker than the electromagnetic interaction, but has a range a hundred times shorter.

- The query on "charge neutrality" indicates whether the charge has both signs, so it can be neutralized. As indicated, gravity is the only force that cannot be "shielded".

The only interactions we can feel in the macroscopic world are the electromagnetic and gravitational interactions, because these are long-ranged. Gravity is weaker than electromagnetism by more than 30 orders of magnitude, but it cannot be shielded, and becomes the dominant force over cosmic distances.

Early phenomenological views of the interactions are represented by the effective Feynman diagrams in Fig. 13.9. As we shall see in

---

[2]Number conservation means a particle can be created or annihilated only in conjunction with its antiparticle, so that the number of particles minus the number of antiparticles is a constant.

Chap. 17, they are now described by gauge theories, and mediated by spin-1 gauge photons. Gravity remains the exception, where the mediating particle has spin 2, because the gauge symmetry is not internal symmetry but Lorentz invariance itself.

14

The Fall of Parity

14.1. Dawn of the post-modern era

The post-modern era of particle physics began in 1957, with the discovery that left and right are basically different. It was a shot across the bow signaling the emergence of a strange new world.

Atomic states have definite *parity*. This means that the mirror image of the wave function, obtained by reversing the sign of all coordinates, is the same wave function apart from a sign. The plus sign corresponds to even parity, and the negative sign, odd parity.

Parity is conserved in atomic transitions. The quantum jumps with photon emission are governed by *selection rules*: the parity of the atomic state must change if the photon emitted has odd parity, and it must not change if the photon has even parity. All this originates from the fact that the electromagnetic interaction is invariant under reflection, and therefore conserves parity.

Parity is also observed to be conserved by the strong interaction.

It had been generally assumed that all interactions conserve parity. For, if nature were not left–right symmetric, what determines the preference for left or right? The kind of what-else-can-it-be argument is not foolproof, for nature has a way of telling you what it can be. And this case, nature speaks.

It all started with the so-called "tau–theta puzzle". A meson called tau was observed to decay into two pions, while one called theta decayed into three pions. Now, the pion is known to have negative intrinsic parity, and hence the 2-pion and 3-pion states have opposite parity. Therefore, so the reasoning went, the tau and the theta must be different particles. The puzzle was that they had exactly the same mass, as far as one could tell. This apparent "parity-doubling" spawned contrived explanations, but none seemed satisfactory.

An obvious explanation was that parity is not conserved; but what about all those alleged experiments verifying its sanctity?

Chen-Ning Yang and Tsung-Dao Lee made a careful review of experimental evidence for parity conservation, and realized that it was lacking for the weak interaction. Accordingly, they proposed experiments to test it. The proposals were generally met with indifference if not ridicule, for most people already knew the answer. Said Wolfgang Pauli:

I cannot believe that God is a weak left-hander.

Eugene Wigner cited the fact that nuclear states have definite parities as indication of parity conservation.[1]

One of the easier ways to test parity conservation in the weak interaction is to examine the spin polarization of the decay products in pion decay: $\pi \to \mu + \bar{\nu}$. If parity is conserved, then μ must have equal probability of being right- or left-handed. T. D. Lee suggested the experiment to his colleague at Columbia University, Leon Lederman, who reportedly laughed it off.

On the other hand, another colleague of Lee's at Columbia, Chien-Shiung Wu, undertook the test in a more difficult experiment, the beta decay of Co^{60}. Again, the objective was to measure the spin polarization of decay products, to see whether the left and right senses occur with the same rate.

---

[1]Wigner mentioned this in 1956 during afternoon tea at the Institute for Advance Study, Princeton, as the author recalls. There were murmurs of agreement among those present except Robert Oppenheimer, who gave the prescient rejoiner, "But, what about the neutrino?"

Fig. 14.1 Upper left: Chien-Shiung Wu (1912–1997) led the experiment that discovered parity violation. Upper right: Tsung-Dao Lee (1926–), who, together with Chen-Ning Yang, suggested the experiment, and explained it in terms of the two-component neutrino. Bottom: Lee and Yang in 1957 at the Institute for Advanced Study, Princeton, possibly discussing parity violation.

In a collaboration at the National Bureau of Standards, Wu reached the verdict after eight months of hard work: parity was violated to the maximal degree possible.[2]

---

[2]Yang and Lee received the Nobel Prize in 1957 for suggesting parity violation; but C. S. Wu, who demonstrated it experimentally, did not shared the honor. This act of omission, like the earlier lockout of Lise Meitner from the prize for nuclear fission, shows that, like all human institutions, the Nobel Prize is not immune from social and cultural bias.

Lee recalled[3] that he was awakened by a telephone call from Wu at around 6 AM on a Friday in 1956, informing him of the fall of parity. He immediately relayed the news to Lederman, who rushed to Columbia University's Nevis Cyclotron with Richard Garwin, set up and completed the pion decay experiment over the weekend. They confirmed the maximal violation of parity conservation.

14.2. Neutrino: a left-handed screw

Anticipating parity violation, Lee and Yang had an explanation ready, wrote a paper, and stashed it away in a drawer. They anticipated violation to the maximal degree, because that had a simple and elegant explanation, namely that the neutrino has intrinsic handedness, like a screw. An experiment by Maurice Goldhaber, Lee Grodzins, and Andrew Sunyar in December, 1957, determined the neutrino to be a left-handed screw.

The neutrino, being a massless spin-1/2 particle, always moves with the velocity of light, with spin pointing either along or opposite the direction of motion. The former corresponds to positive helicity, while the latter corresponds to negative helicity. The helicity describes which way the particle "turns" as it advances, and therefore corresponds to handedness: positive helicity makes a right-handed screw, and negative helicity, left-handed screw.

A massive particle like the electron does not have definite helicity, for it can flip over to the other sign. The helicity becomes "unflippable" only in the limit of infinite momentum, when the mass becomes negligible.

Dirac's relativistic equation requires that the wave function of a massive spin-1/2 particle have four components, corresponding to the spin doublet and the particle–antiparticle duality. For a massless particle, they break up into two independent groups corresponding to right- and left-handed particles.

---

[3]Private communication.

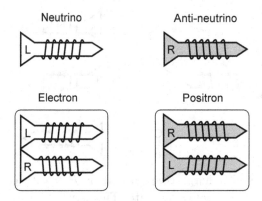

Fig. 14.2 The massless Weyl neutrinos are represented as screws. Anti-neutrinos, shown shaded, have opposite handedness. The physical neutrino is left-handed, while the corresponding antineutrino is right-handed. A right-handed neutrino does not exist. An electron has mass, and must contain both handedness.

These two-component Dirac particles are called "Weyl neutrinos", which were discussed by Hermann Weyl long ago. They are embodiments of right-handed (R) and left-handed (L) screws. The antiparticle, defined as a hole in the Dirac sea, has the opposite handedness from the particle.

The physical neutrino is L, and the antineutrino is R. The electron, being massive, is composed of both R and L. The physical particles are schematically represented in Fig. 14.2.

14.3. CP

We can perform the operations listed below on a Weyl neutrino:

Symbol	Name	Operation	Result
P	Parity conjugation	Spatial reflection	L $\rightleftarrows$ R
C	Charge conjugation	Part.–antipart. exchange	L $\rightleftarrows$ R
CP	CP conjugation	P followed by C	Unchanged

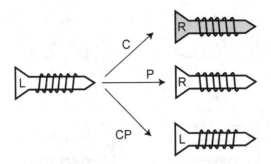

Fig. 14.3 Under C (charge conjugation) or P (parity conjugation), a neutrino goes into a non-physical state. This is why the weak interaction violates C and P separately. It is invariant under CP, however, because the neutrino goes into itself under this operation.

Either P or C takes the physical neutrino to a non-physical state, but CP leaves it intact. This is the reason why the weak interaction violates P and C to the maximal degree, but should conserve CP. This is illustrated in Fig. 14.3.

One might find consolation in the fact that there is symmetry after all. We just have to replace P with CP.

14.4. Is nothing sacred?

But CP is violated, due to an unknown interaction.

Racing ahead of our story, there are "strange" mesons, K^0, $\bar{K}^0$, which are antiparticle to each other. They can be produced by the strong interaction, but they decay via the weak interaction only. The fascinating thing is, the particle that decays is not the particle created, but only "part of it". Neither K^0 nor $\bar{K}^0$ has a definite lifetime. The states that decay with definite lifetimes are the linear super-positions

$$K_{\mathrm{L}} = \frac{1}{\sqrt{2}} \left(K^0 - \bar{K}^0 \right),$$

$$K_{\mathrm{S}} = \frac{1}{\sqrt{2}} \left(K^0 + \bar{K}^0 \right).$$

James W. Cronin   Val L. Fitch   Masatoshi Koshiba
(1931–)   (1923–)   (1926–)

Fig. 14.4 Disturbing the order: CP violation (Cronin and Fitch, 1964); neutrino mass (Koshiba, 1998).

The long-lived state (L) has a lifetime 500 times that of the short-lived state (S). These states have definite signature with respect to CP: K_L is odd, while K_S is even. Thus, K_L should be forbidden to decay into two pions ($\pi^+\pi^-$), which is even under CP.

In 1964, James Cronin and Val Fitch found that K_L, which normally decays into three pions, does decay into two pions, with a branching ratio of 2×10^{-3}. It is a rare decay, but CP is violated.

Fast-forward to 1998. An experiment by the Super Kamiokande group in Japan, led by Masatoshi Koshiba (1926–), demonstrated that the neutrino has mass. The experiment detected a mass difference between two types of neutrinos in the range 0.03–0.1 eV. This represents an extremely small energy, between 3–10% of the energy of atoms. But it signifies that the Weyl neutrino has fallen.

We do not yet understand these small deviations from an otherwise "pretty" picture.

15

The Particle Explosion

15.1. The accelerator boom

To probe the structure of matter to ever smaller length scale, one needs ever higher energy. The energy scale and corresponding distance scale in physics are indicated in the following display:

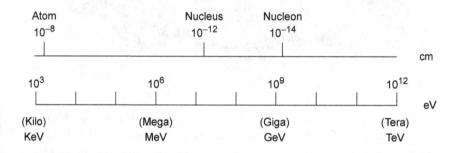

To be an effective probe, the energy must be concentrated in a single subatomic particle, so it could initiate reactions by colliding with another particle. We then try to deduce the inner structure of the particle by studying the reaction products. Feynman likened the process to banging two Swiss watches together and seeing what comes out. But we do learn about the screws that went into those watches.

The high-energy projectiles have to be energized in particle accelerators. From the early 1930's, America has always led in the building of ever more powerful accelerators. Some early ones are shown in Fig. 15.1.

Fig. 15.1 Early accelerators and energies achieved. Clockwise: Van de Graff generator, MIT site at Roundhill, Massachusetts (1931), 1 MeV; Stanley Livingston and Ernest Lawrence at Cyclotron invented by the latter at UC Berkeley (1932), 1.5 MeV; Cosmotron at Brookhaven National Laboratory, Long Island, NY (1953), 3.3 GeV.

In the post World War II era, a grateful America gave physicists all the support they wanted for bigger and better machines, and there began a construction boom. The public funding of such "spiritual" quests may be likened to the building of cathedrals in Europe's Middle-Ages, even though funding agencies may not see it that way. Figure 15.2 showed some large accelerators at SLAC, and Fermilab, and the one under construction at CERN. Going hand-in-hand with accelerator technology are particle detectors, some of them shown in Fig. 15.3.

Fig. 15.2 Cathedrals of our time: large particle accelerators. Left: SLAC (Stanford Linear Accelerator) near San Francisco, a 3 km perpendicular bisector of the San Andreas fault, accelerates electrons to 50 GeV. Right top: booster and main ring of Tevatron collider at Fermilab near Chicago. It accelerates protons to 1 TeV. At one time, areas within the rings were planted with prairie grass kept trim by a herd of buffaloes. Right bottom: white circle marks position of underground accelerator ring, 8.5 km in diameter, of the Large Hadron Collider (LHC) at CERN (Conseil Européen pour la Recherche Nucléare), Geneva, Switzerland. When completed in 2007, it will accelerate protons to 7 TeV. The Jura mountain range looms at top of picture.

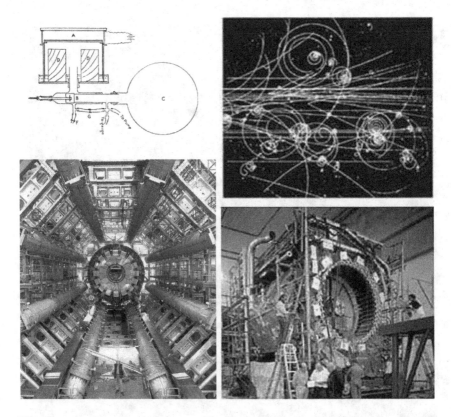

Fig. 15.3 Particle detection. Clockwise: Wilson cloud chamber (1895); particle tracks in bubble chamber (circa 1960); detector Mark II at SLAC (1987); detector ATLAS at CERN (2005).

There was an exponential increase in the accelerator energy as a function of time, as we can see from the "Livingston plot" in Fig. 15.4.

15.2. Darkness at noon

A surge of new particles began around 1950, as chronicled in Fig. 15.5. They came so fast and thick that one experimentalist remarked:

> You used to get a Nobel Prize for discovering a particle. Now you should be fined ten thousand dollars.

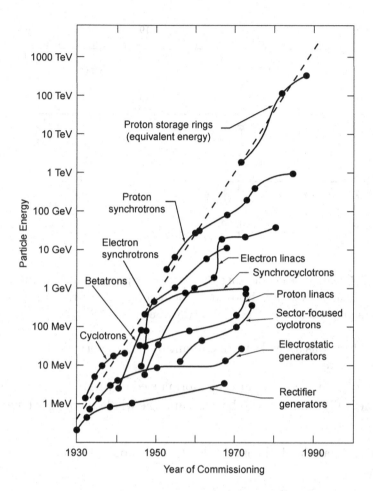

Fig. 15.4 Livingston plot: energy of accelerators as function of time. From W. K. H. Panofsky, *Beamline* 27-1, p. 36. (www.slac.stanford.edu/pubs/beamline/27/1/27-1-panofsky.pdf)

Victor Weisskopf, a theorist, lamented:

You spend millions to create a particle; then you have to explain it.

In contrast to experiment, theory went into a depression after the blinding success of QED. The perturbation theory that had worked so well was useless for the new physics, for it involved the strong interaction, with a coupling strength of 1 instead of 1/137 as in

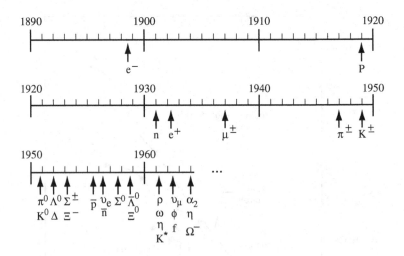

Fig. 15.5 Time scale showing the particle explosion since about 1950. Names of particles appear below dates of discovery. After 1965 the particle were too numerous to be shown this way. (See Fig. 15.8.)

QED. There was a revolt against quantum field theory, perhaps out of disillusionment.

Lev Landau argued that the renormalized charge of the electron in QED actually comes out to be zero, and the whole theory is "trivial". He declared that

> Hamiltonian field theory is dead, and should be buried with honors.

Steven Weinberg simply said

> The market has crashed.

15.3. The ontological bootstrap

There appeared a movement to treat all particles on equal footing, instead of assigning fundamental status to a few as in quantum field theory. It was prompted by the explosion of particles on the experiment front, and, on the theory front, by Tullio Regge's theory of recurrent resonances of ever higher spin.

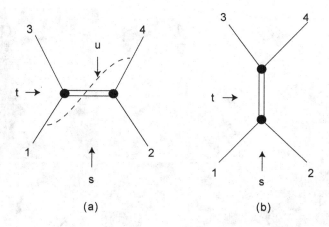

Fig. 15.6   The bootstrap: (a) Interaction in the s-channel is due to exchange of resonances in the t-channel. (b) The interaction leads to formation of resonances in the s-channel. Requiring the two set of resonances to be the same determines the spectrum.

Geoffrey Chew, the high priest of the movement, called the approach "nuclear democracy". The philosophy was that the observed particles were bound states of *one another*. The universe "bootstraped" itself into existence.[1]

Consider a collision between two identical particles, as represented by the Feynman diagrams in Fig. 15.6, where the possible collision channels are labeled s, t, u.

The gospel is as follows:

The interaction in the s channel is generated by exchange of resonances (unstable particles) in the crossed channels t and u, and it creates resonances in the s channel. The requirement that the s resonances be identical to the t and u resonances determines their spectrum. In this sense, the resonances "bootstrap" themselves into existence.

---

[1] By curious coincidence, the rebellion against quantum field theory in the 1960's centered in Berkeley, California, side-by-side with the social rebellion of the Flower Children. The Birge Hall of the bootstrap is just blocks away from the Telegraph Avenue of the hippies.

Fig. 15.7 Warriors of the "bootstrap" era. Clockwise: Geoffrey Chew (1924–); Chung-I Tan (1942–), seated, staring at Chew; Sergio Fubini (1928–2005); Tullio Regge (1931–); Stanley Mandelstam (1928–); Gabriele Veneziano (1942–). The bootstrap idea was steamrollered by the discovery of quarks, but lives to fight another day, in a reincarnation called "string theory", popularly known as *The Theory of Everything*.

The mathematical formulation centers on the "S-matrix" of the scattering process. From general principles, it must have two symmetries: unitarity and crossing-symmetry. The former guarantees the conservation of matter, and the latter says that one matrix describes all channels by analytic continuation. These two requirements are at loggerheads, and nearly impossible to reconcile. Chew proclaims that only one S-matrix can satisfy both requirements, and that is the S-matrix of the world.

Gottfried Wilhelm Leibniz (1646–1716) famously argued that our world is the best of all possible worlds. Chew puts it on a higher ontological plane:

Our world is the only possible world.

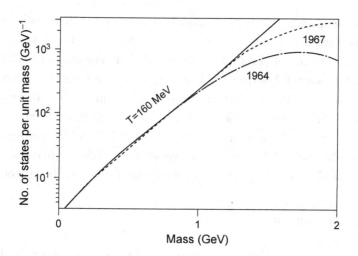

Fig. 15.8 Particle spectrum appears to rise exponentially with mass. Dotted lines indicate experimentally observed particles as of the year indicated. The exponential envelop corresponds to a characteristic energy of the same order as the pion mass. It represents the ultimate temperature that any system can attain.

15.4. The ultimate temperature

In 1968 Gabriele Veneziano wrote down an amplitude that exhibits the bootstrap, with the same set of resonances in the s, t, u channels. Called the *dual resonance model*, the formula was the product of sheer imagination, with no theory behind it. Steven Weinberg[2] commented at the time that the model is "so pretty that it ought to be correct, if there is any justice," and added, "But as we know, there is no justice."

Sergio Fubini, Stanley Mandelstam, and others extended the dual resonance model to multiparticle amplitudes, and found that, for consistency, the number of resonances must grow exponentially with mass. Now *that* is an explosion. As we can see in Fig. 15.8 it appears to be borne out by experiments.

---

[2]Private communication.

If the exponential trend continues indefinitely, there will be an *ultimate temperature*. If you try to pump heat into any system, the temperature will not rise beyond some value. There would be so many new particles available that the heat will create new particles in order to increase the entropy, instead of raising the temperature of existing particles. Various theoretical estimates give an ultimate temperature of the order of the pion mass, around 160 MeV. This temperature was also obtained earlier by Rolf Hagedorn (1919–2003) from particle data. If valid, it has important implications for the early universe.[3]

15.5. Echos of an era

It was soon realized that Veneziano's dual resonances can arise from the vibrations of a "dual string". But work along these lines stalled when quarks were discovered. People felt relieved that something was fundamental after all, went back to quantum field theory, and resurrected Yang–Mills gauge theory.

The dual string was born again later on the Planck scale, a length smaller than the nucleon radius by a factor of 10^{-19}. The new theory hangs on the hope of unifying quantum gravity with other interactions, and is known in the media as *The Theory of Everything*.

---

[3]K. Huang and S. Weinberg, *Phys. Rev. Lett.* **25**, 895 (1970).

16

Quarks

16.1. Strangeness

A key to the classification of particles is to identify conserved quantum numbers. It has been known for a long time that the strong interaction conserves isotopic spin. A new quantum number now enters the picture: "strangeness".

Some particles are produced in accelerators at a high rate, but once created, they decay exceedingly slowly. For example, a baryon called Λ^0 can be created in a collision between π and p, and it decays back to π and p; but the decay rate is smaller than the production rate by a factor of 10^{-13}.

Murray Gell-Mann and Abraham Pais offered an explanation of this striking phenomenon in terms of "associated production", namely, Λ^0 was produced in the company of a meson K^0, whereas it decays alone. The high transition rate requires the participation of a cohort.

Gell-Mann made it concrete by proposing an additive quantum number called "strangeness", which is conserved by the strong interaction, but violated by the weak interaction. Ordinary particles are assigned zero strangeness, while the partners in associated production, Λ^0 and K^0, are assigned strangeness $+1$ and -1 respectively. The pair Λ^0–K^0 has zero total strangeness, and thus can be created via the strong interaction in the process

$$\pi + p \rightarrow \Lambda^0 + K^0 \, .$$

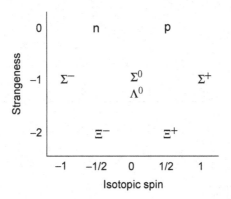

Fig. 16.1   The nucleon is a member of an octet of spin $1/2$ baryons. A plot of strangness versus isotopic spin produces the hexagonal pattern shown. Octets of spin 0 and and spin 1 mesons are also observed.

However, the total strangeness must change in the decay

$$\Lambda^0 \longrightarrow \pi + p,$$

and it goes via the weak interaction. The difference in the production and decay rates just reflects the difference in strength between the strong and weak interactions.

The new quantum number was also proposed, independently, by Tadao Nakano and Kazuhiko Nishijima around 1953.

16.2. Octet and decuplet

Striking patterns emerge when hadrons are sorted according to isotopic spin and strangeness. For example, the familiar nucleon doublet appears in a family of eight baryons all with spin $1/2$, as shown in Fig. 16.1. Isotopic spin (I) multiplets are displayed horizontally, and different rows have different strangeness. The nucleon doublet with $I = 1/2$ sits at top, followed by the Σ triplet with $I = 1$, and the Ξ doublet with $I = 1/2$. The singlet Λ^0 sits at the center. This is called the nucleon octet. There are other octets: the π octet consisting of spin 0 mesons, and the ρ octet of spin-1 vector mesons. These are indicated in Fig. 16.2.

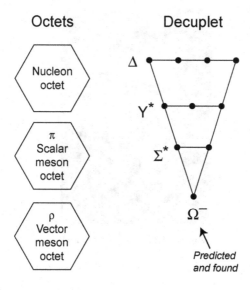

Fig. 16.2 In the eightfold way, hadrons are organized into octets and a decuplet. These groupings are representations of the group $SU(3)$, which has 8 generators. The last member in the decuplet, Ω^-, was discovered after Gell-Mann predicted it.

In the early 1950's, Fermi's team at Chicago discovered a baryon that caused a stir. It occurred as a prominent resonance in π-p scattering with spin 3/2 and isotopic spin 3/2. Called the "three-three resonance", it was regarded as an excited state of the nucleon, and key to its inner structure. Furious theoretical activity to explain it ran into blank walls. Fallen by the wayside were such relics as "strong-coupling theory" and "Tamm–Dancoff approximation". A phenomenological theory of Geoffrey Chew and Francis Low scored a small success, and eventually led Chew to his bootstrap model.

Dyson expressed his frustration by saying, "We wouldn't understand it in a hundred years." Said Fermi, "I probably will not understand it in my lifetime." That was sadly prophetic, for he died at age 51 in 1954.[1]

---

[1] Dyson and Fermi made the remarks in their Loeb Lectures at Harvard University, around 1953.

Murray Gell-Mann (1929–)

Fig. 16.3 Murray Gell-Mann created strangeness, perceived the eightfold way, and found quarks.

With the discovery of other baryons, the three-three resonance is renamed Δ, and fits into a family of ten, as shown in Fig. 16.2. It does not seem particularly special, nor directly relevant to nucleon structure, which is due to quarks, as we shall see.

16.3. The eightfold way

Gell-Mann and Yuval Ne'eman recognized that the octet and decuplet are representations of the group $SU(3)$, which has 8 generators. The octet is the adjoint representation. With a flair for terminology, Gell-Mann calls it the "eightfold way", after a Buddhist teaching.

The eightfold way implies that $SU(3)$ is an internal symmetry group of the strong interaction. But the symmetry is only approximate, for, despite the recognizable patterns, particles in the octets and the decuplet do not have the same mass. Gell-Mann, and independently, Susumu Okubo, proposed a specific manner in which the symmetry is violated. This led to a mass formula that predicted an equal spacing of masses in the Δ decuplet.

Nicholas Samios (1936-)

Fig. 16.4 The last piece fell into place: in 1964, the Omega minus was discovered at the AGS (Alternating Gradient Synchrotron) at Brookhaven National Laboratory. Clockwise from left: team leader Nicholas Samios; the 60″ bubble chamber used for detection; photograph of track in the reaction recorded; diagram of tracks, in which the Omega minus is seen at lower left, just above the incident K minus.

The last member in the decuplet Ω^- had not been discovered experimentally. With the mass formula, all properties of the missing particle were predicted, and it was found in 1964 by a group led by Nicholas Samios at Brookhaven National Laboratories. (See Fig. 16.4.) This is the subatomic analog of the discovery of the planet Neptune more than a century earlier, and firmly establishes the eight-fold way.

16.4. "Three quarks for Muster Mark!"

Having found the octet and decuplet representations of $SU(3)$, one cannot help but ask, "where is the fundamental representation?" It would be realized by a triplet of particles. Gell-Mann, and independently, George Zweig, pointed out that members of the triplet will have charges quantized not in units of the electronic charge, but a third of that.

Since it would take three of these fundamental particles to make a proton, Gell-Mann called them "quarks". The name sprang from his erudite mind, in free association on a poem in James Joyce's *Finnegan's Wake*:

> Three quarks for Muster Mark!
> Sure he hasn't got much of a bark
> And sure any he has it's all beside the mark.

The three quarks consist of an isotopic spin doublet denoted by u (up), d (down), and a strange singlet s. In the idealized world in which the eightfold way is an exact symmetry, these quarks would have the same mass. In the real world, their mass differences give rise to the mass formula of Gell-Mann and Okubo. The u–d mass difference should be smaller than the u–s mass difference, since isotopic spin is better conserved than strangeness.

The quarks have spin 1/2, baron number 1/3, and properties summarized below:

Quark		Isotopic spin	Strangeness	Charge/e
u	up	1/2	0	2/3
d	down	−1/2	0	−1/3
s	strange	0	−1	−1/3

The eightfold way is automatically implemented when we regard hadrons as bound states of quarks. For example, a proton is $\{uud\}$, and a neutron is $\{ddu\}$. The quark types u, d, s are referred to as "flavors".

We have yet to introduce another quantum number — "color". There are to be three colors, and each of the three quarks in the proton have different colors. Quark dynamics will spring from color.

16.5. Charm and beyond

More quark flavors appeared. The first extra flavor is "charm", proposed in 1970 by Sheldon Glashow, John Illiopoulos, and Luciano Maiani, for consistency in the theory of the weak interaction. Two more were proposed later, to supply missing members from a compelling family picture, and were named "top" and "bottom".[2]

With these, the number of quark flavors balloons from 3 to 6, and all have been established experimentally. We summarize their properties in the following table:

Quark		Flavor	Charge/e	Mass (MeV)
u	up	$I_3 = +1/2$	$+2/3$	1.5–4.0
d	down	$I_3 = -1/2$	$-1/3$	4–8
s	strange	$S = -1$	$-1/3$	80–130
c	charm	$C = 1$	$+2/3$	1150–1350
b	bottom	$B = -1$	$-1/3$	4100–4400
t	top	$T = 1$	$+2/3$	71400 ± 2100

The masses are not measurements, but parameters used in theoretical models. The c, b, t quarks are called "heavy quarks".

16.6. Partons

A hundred years ago, Rutherford scattered α-particles from atoms, and, from the prevalence of backscattering, concluded that there is something small and hard inside batting back the incoming projectiles. The atomic nucleus was thus discovered.

---

[2]In Europe, some started to call the two new quarks "truth" and "beauty". Mercifully, that didn't wash.

James D. Bjorken   Jerome I. Friedman   Henry W. Kendall   Richard E. Taylor
(1934–)   (1930–)   (1926–1999)   (1929–)

Fig. 16.5   A modern Rutherford experiment: Scattering of electrons from a proton reveals that there are point-like scatterers inside. Bjorken laid the theoretical basis for analysis of the experiment led by Friedman, Kendall, and Taylor. These scatterers have come to be identified with quarks.

A modern version of the Rutherford experiment was performed in 1968, by a team from MIT and Stanford University, led by Jerome Friedman, Henry Kendall, and Richard Taylor. They scattered electrons from a proton, and, analyzing the data using a theory by James Bjorken, concluded that there are point-like scatterers inside.

At the height of the "nuclear democracy" movement, they found that the proton is not made from molasses bootstrapping itself into existence. There's grit inside; the proton has "parts". In his straight-shooting way, Feynman called them "partons".

People began to think they might be quarks. Since quarks are spin-1/2 objects, they should obey the Dirac equation, and have a Dirac sea in the vacuum. The partons may just be the quarks of the eightfold way, plus swarms of quarks and antiquarks excited from the Dirac sea.

The clincher came six years later.

16.7. Charmonium

In the annals of experimental physics, some feats stand out as epoch-making, in that they change people's perception's almost overnight.

Samuel C. C. Ting   Burton Richter
(1936–)   (1931–)

Fig. 16.6 Ting named his particle J, and Richter named it ψ, and so it is known as J/ψ. Structurally it is charmonium — a bound state of the charmed quark and its antiparticle.

One such experiment was Perrin's measurement of Avogadro's number from Brownian motion, which demonstrated the reality of atoms. Another was the fall of parity, revealing a fundamental difference between left and right. It happened again in November 1974, with the discovery of a particle named J/ψ. Called the "November revolution", it established the reality of quarks.

The J/ψ is a vector meson of mass of 3.1 GeV — more than three times that of the proton. Its lifetime however, is a thousand times longer than ordinary unstable hadrons. In Ting's words, discovering this particle was like

> stumbling upon a village inhabited by people who live to be ten thousand.

In the experimental data, the particle appears as a peak in the yield of electron–positron scattering. The long lifetime means that the peak is extremely sharp — the proverbial needle in a haystack.

It was soon recognized that the J/ψ is charmonium, the bound state of a charm and anticharm quark. The charmed quark is so heavy that charmonium can be described by non-relativistic quantum mechanics, with a spectrum of excited states as shown in Fig. 16.7. The J/ψ may be said to be the "hydrogen atom" of quark physics.

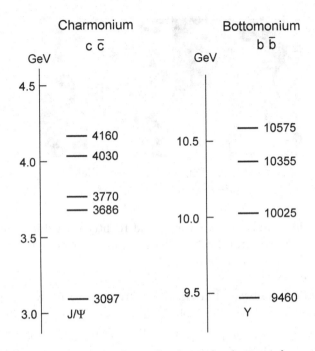

Fig. 16.7 The "hydrogens atoms" of quark physics: charmonium ($c\bar{c}$) and bottomonium ($b\bar{b}$). The quarks are so heavy that the spectra can be calculated using non-relavistic quantum mechanics. Each level represents a vector meson.

Bottomonuim — bound state of bottom and antibottom quarks — was observed in 1977, and named Υ (Upsilon). Its spectrum of excited states is also shown in Fig. 16.7.

16.8. Color

Once quarks are accepted as real, it is natural to regard the nucleon as a dynamical bound state of three quarks. In a simple model, one puts the quarks into orbitals in a central potential, like electrons in an atom. Experiments show that the magnetic moment of a nucleon is close to a sum of quark magnetic moments. This suggests that all three quarks are in the lowest orbital; but this is impossible for they have spin 1/2, and should obey the Pauli exclusion principle. The way out is to endow them with a new attribute, so the three quarks are not identical.

Since we need to put three quarks into the same state, each quark flavor should come in three copies. Again, Gell-Mann says it with pizazz: each flavor comes in three different "colors". He called them *red*, *white*, and *blue* originally, but the conventional now is *red, yellow* and *green*.

Unlike flavor, color is an exact symmetry of the strong interaction. Since there are three colors, the symmetry group is color $SU(3)$, not to be confused with flavor $SU(3)$.

A nucleon is then made up three quarks, each of a different color. The nucleon contains an equal mixture of the primary colors, and is "colorless". Mathematically this means that the nucleon is a color singlet — it is invariant under color $SU(3)$.

We can now formulate a rule to explain why quarks have not been directly observed:

Only colorless states can physically exist.

This dictum is known as "quark confinement", or more accurately, "color confinement". We shall explain how it comes about in Chap. 19.

The exactness of the color symmetry means that it can be gauged. The resulting Yang–Mills theory is QCD (quantum chromodynamics), the theory of the strong interaction.

17

All Interactions are Local

17.1. Yang–Mills awakens

Thomas "Tip" O'Neill (1912–1994), late speaker of the U.S. Congress from Massachusetts, declared that "All politics is local." In elementary particle physics, all interactions are local. The reason is the same: there is no such thing as action-at-a-distance.

After living with quarks and leptons for a while, people began to entertain the idea that their interactions might be generated by some gauge principle. After slumbering for almost twenty years, Yang–Mills theory was called to service, to begin the construction of the gauge theory of non-gravitational interactions.

The steps are as follows:

- Start with free leptons and quarks, which are Weyl neutrinos described by Dirac theory.
- Identify the global symmetry to be gauged, by stating the gauge groups and the representations.
- Introduce scalar fields needed to generate mass in a gauge-invariant manner.
- Introduce the necessary gauge fields to promote global gauge invariance to local gauge invariance. The gauge couplings appear through the covariant derivatives.

The basic leptons and quarks are Weyl neutrinos, with the left-hand screws and right-handed screws regarded as independent massless particles. Mass will be a derived quantity emerging from gauge-invariant couplings to scalar fields, as we will explain later.

The physical electron, for example, consists of a left-handed screw and right-handed screw held together with a mass term. The screws involved belong to different representations of the gauge group, and are really independent particles.

As is now known, physical neutrinos have small masses. They are therefore composed from Weyl neutrinos, like the electron. For simplicity, however, we pretend that neutrinos are massless.

In the following, we first describe how to unify the electromagnetic and weak interactions, in a model with only electron e and neutrino ν. When the principle of the theory is made clear, we shall add the u and d quarks, which together with e, ν, make up a close-knit family. We then introduce QCD, the gauge theory of the strong interaction of the quarks.

Finally, we take into account two other lepton-quark families to complete the so-called *Standard Model* of particle physics.

In the end, we shall have a model based on a single principle — the gauge principle — and it has withstood confrontation with experiments. However, it is very intricate, containing a large number of constituents and empirical coefficients.

Many intriguing aspects of the Standard Model deserve to be examined in greater detail, and we shall describe them from a physical point of view, in separate chapters:

- Spontaneous symmetry breaking (Chap. 18).
- Quark confinement (Chap. 19).
- Triangle anomaly (Chap 20).

17.2. Unifying electromagnetic and weak interactions

We first describe the global symmetry to be gauged, in the unified theory of weak and electromagnetic interactions.

The electron e and neutrino ν are built from right- and left-handed screws, the Weyl neutrinos R and L. The physical ν is pure L, while e contains both L and R, tied together to generate mass.

We start with three basic particles: ν_L, e_L, e_R, where the subscripts label their handedness. The two Ls are equivalent. That is, the Hamiltonian should be invariant under a group $SU(2)$. The doublet $\{\nu_L, e_L\}$ forms a fundamental representation of this group, while e_R is an invariant singlet. This is indicated in the table below:

$$\text{Doublet:} \quad \begin{pmatrix} \nu_L \\ e_L \end{pmatrix},$$

$$\text{Singlet:} \quad (e_R).$$

Since we aim to generate the electromagnetic interaction, electric charge must enter the picture. Assume that the Hamiltonian is invariant under independent phase changes of the three screws. There are two independent relative phases. One of these corresponds to "lepton number",[1] which will not be gauged. The other phase contains the charge, and represents a $U(1)$ transformation. The symmetry group to be gauged is therefore $SU(2) \times U(1)$.

Let us denote the generators of $SU(2)$ by L_1, L_2, L_3, and the generator of $U(1)$ by L_0. The electric charge is defined as

$$Q = L_0 + L_3.$$

The $SU(2)$ generators are represented as follows:

$$L_a = \begin{cases} \sigma_a/2 & \text{(Doublet representation)} \\ 0 & \text{(Singlet representation)} \end{cases} \quad (a = 1, 2, 3).$$

Now we turn on the interaction by gauging $SU(2) \times U(1)$. We need to introduce one gauge field for each generator, and so there are four gauge fields

$$W_1, W_2, W_3, W_0,$$

each of which is a 4-vector. The ordinary derivative ∂ is replaced by the covariant derivative in the following manner:

$$\partial \rightarrow \partial + ig(L_1 W_1 + L_2 W_2 + L_3 W_3) + ig' L_0 W_0,$$

---

[1]Lepton number is a conserved quantity, such that the number of leptons minus the number of antileptons is a constant.

where g and g' are two coupling constants corresponding to the gauge groups $SU(2)$ and $U(1)$, respectively.

The leptons now can emit or absorb the vector gauge photons. The charges can be read from the covariant derivative:

Group	Gauge fields	Charges
$SU(2)$	W_1, W_2, W_3	gL_1, gL_2, gL_3
$U(1)$	W_0	$g'L_0$

The right-handed e_R does not interact with W_1, W_2, W_3, because the corresponding generators are represented as zero.

17.3. Generating mass

Mass appears in the Hamiltonian as the coefficient of a term that is a product of L and R. But L changes under $SU(2)$, while R remains invariant. Such a term with constant mass is unacceptable, because it is not gauge invariant.

To get around this, we replace the mass coefficient by a complex scalar field ϕ, which transforms as a doublet under $SU(2)$:

$$\phi = \begin{pmatrix} \phi_+ \\ \phi_0 \end{pmatrix},$$

where ϕ_+ carries positive electric charge, and ϕ_0 is neutral. We can then combine ϕ with e_L to produce an invariant under $SU(2)$. The mass is then proportional to ϕ_0. The complex scalar field ϕ is called the *Higgs field*.

To obtain the observed mass of the electron in the vacuum, we arrange for the field to be non-zero in the vacuum state. This is done by introducing a potential energy that depends on $|\phi|^2$, with a minimum away from zero, as illustrated in Fig. 17.1. This phenomenon is called *spontaneous symmetry breaking*, which will be discussed in more detail in Chap. 18.

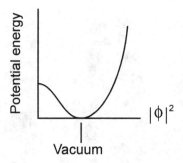

Fig. 17.1 Potential energy for the Higgs field, with a minimum at a non-zero value of the field. This generates masses in a gauge-invariant manner.

17.4. Making the photon

The Higgs field generates mass for the gauge field, through the action of the covariant derivative on it. This removes the difficulty with massless gauge photons. But one gauge photon had better remain massless — the physical photon.

The electromagnetic field A is the gauge field that is coupled to the charge $Q = L_0 + L_3$. To find out what it is in terms of the W gauge fields, we go back to re-examine the covariant derivative. It turns out the photon is a combination of W_0 and W_3. To display it, make a rotation in the W_0-W_3 plane to obtain two new gauge fields:

$$A = W_0 \cos \theta_w + W_3 \sin \theta_w ,$$

$$Z = -W_0 \sin \theta_w + W_3 \cos \theta_w .$$

The angle θ_w is called the *Weinberg angle*, so chosen that A is multiplied by Q in the covariant derivative. This imposes relations between the gauge couplings g, g'. The electromagnetic field A has zero mass, because of the way charge is defined.

The Weinberg angle can be measured experimentally, and is found to correspond to $\sin^2 \theta_w = 0.2$.

The gauge fields W_0, W_1, W_2, W_3 can now be reorganized into

$$W^+, W^-, Z, A,$$

Sheldon Glashow   Steven Weinberg   Abdus Salam
(1932–)   (1933–)   (1926–1996)

Fig. 17.2 Glashow proposed the symmetry $SU(2) \times U(1)$ for the electroweak sector. Weinberg and Salam gauged it, and introduces a Higg field to generate mass.

where W^+, W^- are linear combinations of W_1, W_2 that possess equal and opposite electric charge. With the Weinberg angle and coupling constants taken from experiments, the theory predicts the mass of $W^\pm$ and Z:

$$m_W = 80 \text{ GeV},$$
$$m_Z = 90 \text{ GeV}.$$

These heavy gauge photons are nearly a hundred times heavier the proton, and makes the range of the weak force 10^{-15} cm, a hundred times shorter than the nuclear scale of 10^{-13} cm.

17.5. Historical note

Historically, the gauge theory of the unified electroweak interaction was the work of Sheldon Glashow, Steven Weinberg, and Abdus Salam, carried out independently over a number of years in the 1960's. Glashow proposed the $SU(2) \times U(1)$ group; Weinberg and Salam gauged the symmetry and introduced the Higgs field. In 1971 Gerald 't Hooft completed work started by Martinus Veltman to show

Gerard 't Hooft   Martinus J. G. Veltman
(1946–)   (1931–)

Fig. 17.3 The electroweak theory is proven to be renormalizable, and thus can be used for computations.

that the gauge theory is renormalizable. This made it possible to do practical calculations in perturbation theory.

The heavy gauge photons W and Z were discovered experimentally in 1983 at CERN, in a large project led by Carlo Rubbia and Simon van der Meer. Their measured masses agree with theoretical predictions.

17.6. The lepton-quark family

The quarks u, d acquire the electroweak interaction by joining ν, e to form one family $\{e, \nu, u, d\}$, as described by the following updated representation of the weak $SU(2)$ group:

$$\text{Doublets: } \begin{pmatrix} \nu_L \\ e_L \end{pmatrix}, \quad \begin{pmatrix} u_L \\ d_L \end{pmatrix}$$

$$\text{Singlets: } \quad e_R \quad u_R \quad d_R.$$

We must not forget that each of the quarks come in triplicate, corresponding to color. All color copies are coupled in the same manner to the gauge fields and the Higgs field.

Carlo Rubbia   Simon van der Meer
(1934–)   (1925–)

Fig. 17.4   Discovery of W and Z, the heavy gauge photons that mediate electroweak interactions.

17.7. QCD

Quarks are involved in the strong interaction in addition to the electroweak interaction. The former arises by the gauging of color $SU(3)$. We label the colors

red, yellow, green.

A color triplet is represented as a vector in color space:

$$\vec{u} = \begin{pmatrix} u_{\text{red}} \\ u_{\text{yellow}} \\ u_{\text{green}} \end{pmatrix}, \quad \vec{d} = \begin{pmatrix} d_{\text{red}} \\ d_{\text{yellow}} \\ d_{\text{green}} \end{pmatrix}.$$

Each vector forms a fundamental representation of color $SU(3)$. That is, a color transformation mixes the components in each of the vectors above, and the left- and right-handed quark components transform in the same way. The color transformation does not mix flavor.

The color group $SU(3)$ has eight generators t_b ($b = 1, \ldots, 8$). To gauge the color symmetry, we need to introduce eight gauge fields G_b ($b = 1, \ldots, 8$). The gauge photons are called "gluons", since they "glue" quarks together to form hadrons.

The strong interaction is turned on by replacing ordinary derivative with covariant derivative:

$$\partial \to \partial + i\lambda t_b G_b \,,$$

where λ is the gauge coupling constant, and the repeated index b is summed from 1 to 8. The quarks now can emit and absorb gluons. The gluons themselves carry color charge, and they can emit and absorb themselves.

We have now gauged $SU(3) \times SU(2) \times U(1)$, in a system consisting of the lepton-quark family

$$\left\{ \begin{matrix} \nu & e \\ \vec{u} & \vec{d} \end{matrix} \right\}.$$

The electroweak group $SU(2) \times U(1)$ mixes particles horizontally, in both rows. The color group $SU(3)$ mixes components of the vectors $\vec{u}, \vec{d}$ in color space.

The quarks and the gluons have never been observed directly. The explanation is "quark confinement" or "color confinement", namely only color singlet states can exist in isolation (Chap. 19).

The lepton-quark family has a total of 15 Weyl neutrinos. As we shall see in Chap. 20, however, it is really one inseparable particle, for the omission of any component will incur the wrath of unmaskable ultraviolet catastrophe.

17.8. Two more families: who ordered them?

There are two other families similar to the one described above, with the same symmetry properties. One is composed of the muon μ, its own neutrino ν', and the strange and charmed quarks. The discovery of the muon neutrino ν' in 1963 by Leon Lederman, Melvin Schwartz, and Jack Steinberger, was a significant milestone in our understanding of leptons.

In 1975, Martin Perl discovered the τ, a spin-1/2 fermion more than a thousand times heavier than the proton. It only has electromagnetic and weak interactions, and earns the oxymoron "heavy

Fig. 17.5 In 1963 Leon Lederman (1922–), Melvin Schwartz (1932–2006), and Jack Steinberger (1921–) discovered that the muon neutrino is distinct from the electron neutrino.

Martin Lewis Perl (1927–)

Fig. 17.6 In 1975 Martin Perl discovered the τ, a third lepton besides e and μ. With a mass almost twice that of the proton, it is no longer "light", as the root meaning of "lepton" would indicate.

lepton". It is paired with a neutrino ν'', and the new lepton duo is joined by the bottom and top quarks to form the third family.

The existence of the τ neutrino and top quark were long anticipated because of the family structure, but were not found till much later. The top quark was experimentally created in 1995, and the τ in 2000, both at the Fermilab, through the joint effort of many groups of workers.

We have no idea why there is more than one family. We hear Rabi's question, "Who ordered them?"

17.9. The standard model

The three lepton-quark families are displayed in the table below:

$$
\begin{array}{ccc}
\mathrm{I} & \mathrm{II} & \mathrm{III}
\end{array}
$$

$$
\left\{\begin{array}{cc} \nu & e \\ \vec{u} & \vec{d} \end{array}\right\}
\left\{\begin{array}{cc} \nu' & \mu \\ \vec{c} & \vec{s} \end{array}\right\}
\left\{\begin{array}{cc} \nu'' & \tau \\ \vec{t} & \vec{b} \end{array}\right\} .
$$

They are identical in terms of group representation and gauge couplings. They are coupled to the Higgs field with the same general form but different mass parameters.

The entries in the table above are objects that form group representations, but they are not the physical particles. The experimentally observed particles are mixtures of entities across families. This mixing adds a level of intricacy to the theory, not to mention a large number of phenomenological parameters.

In broad outline, we have described the gauge theory of strong and electroweak interactions, known as the *Standard Model*, a name attributed to Steven Weinberg. It is the best quantum blueprint of the world we have, excluding gravitation.

The drab name "Standard Model" pales in comparison with the adventuresome "strangeness", or the exuberant "quark". But it is a fitting designation for a work in progress. With the profusion of building blocks, coupling constants, and masses, it is hard to believe there is not something more basic beneath the surface.

18

Broken Symmetry

18.1. What is mass?

The Standard Model revises our conception of mass.

In Newtonian mechanics mass was an intrinsic attribute of a body. This view is no longer tenable in the gauge theory of particle interactions, for it violates gauge invariance. Instead, mass is a property like the magnetism of a ferromagnet: it appears in a certain thermodynamic phase, and can disappear in a phase transition.

The basic players in the Standard Model are massless Weyl neutrinos interacting via gauge couplings. To generate mass in a gauge-invariant manner, the gauge symmetry must be broken — not explicitly but "spontaneously". This means that, while the Hamiltonian continues to be invariant under a gauge transformation, the ground state of the system is not invariant. This is accomplished by introducing the Higgs field, which does not vanish in the vacuum state. Then,

- particle masses arise from mass terms in which the Higgs field appears where the mass was supposed to be;
- gauge photons acquire mass through the covariant derivative of the Higgs field.

The vacuum value of the Higgs field that generates mass depends on the effective potential energy, which is put into the model "by hand".

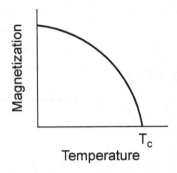

Fig. 18.1 Ferromagnetic phase transition: the magnetiztion is zero below a critical temperature, and grows with decreasing temperature below that point.

What we have here is a specific application of a very general phenomenon — spontaneous symmetry breaking. It underlies phase transitions, notably ferromagnetism and superconductivity. We shall approach the subject from a general perspective, with emphasis on the physical picture.

18.2. How a magnet gets magnetized

A ferromagnet loses its magnetization above a certain critical temperature, but regains it when recooled. The process is a reversible thermodynamic phase transition, as illustrated in Fig. 18.1. The underlying cause of this phenomenon is that atoms of the material have spin, with corresponding magnetic moment. Magnetic interactions favor alignment of the spins, while thermal fluctuations tend to randomize their directions. These two opposing tendencies compete for dominance, and the outcome depends on the temperature. Above the critical temperature, thermal fluctuation wins, and there is no net magnetization. Below that temperature, interaction wins, and the system becomes one big magnet.

In a volume inside the macroscopic system, far away from boundaries, the system has no preferred direction in space, i.e. the Hamiltonian is invariant under rotations. When it magnetizes, however,

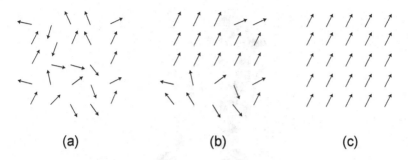

(a)   **(b)**   **(c)**

Fig. 18.2 Development of spontaneous magnetization as the temperature decreases. (a) At high temperature, atomic spins have random orientations. (b) As the temperature decreases, spins align to form local clusters, but the total spin still averages to zero. (c) At the critical temperature, an avalanche occurs towards one big cluster.

the total magnetic moment must point along some direction in space. We say that the system breaks the rotational invariance "spontaneously".

How does this happen? What determines the direction of the magnetization?

As the magnet cools from a high temperature, the spin distribution changes as illustrated in Fig. 18.2. Initially the spins were randomly oriented in space. When the temperature is decreased, clusters of aligned spins occur, but the overall magnetization is still close to zero. When the temperature drops to a certain critical value, there is an avalanche into a single cluster. The final direction of the magnetization is randomly chosen, being that of the cluster which started the avalanche. The avalanche is faster for a larger system, and in the limit of an infinite system it happens suddenly.

Thermal noise is ever present, and the directions of individual spins fluctuate. At high temperatures individual spins fluctuate independently, and the system samples all possible spin distributions in a short time. When clusters start to form however, it takes longer to sample those states corresponding to rotations of the cluster as a whole. That requires synchronized motion of a large number of spins, and rarely happens by chance. The bigger the cluster, the rarer it

Lev Davidovich Landau (1908–1968)

Fig. 18.3   Landau introduces the order parameter to describe a phase transition.

becomes, and the lifetime of a cluster increases exponentially with the number of spins. For a magnetic with the order of 10^{23} spins in a single cluster, this lifetime is overwhelmingly larger than the age of the universe, and therefore rotational symmetry appears to be broken.

In summary,

> Spontaneous symmetry breaking occurs because the system gets stuck in a pocket of biased states, and does not have sufficient time to sample the balancing states. It signifies a failure of ergodicity.

18.3.   The order parameter

Lev Landau proposed a phenomenological description of spontaneous symmetry breaking, independent of the detailed mechanism that causes it. It is based on the idea of the "order parameter", prototype of the Higgs field.

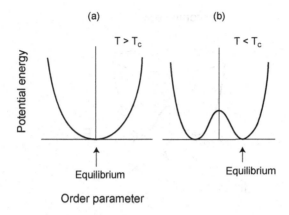

Fig. 18.4 In equilibrium, the order parameter sits at a minimum of the potential energy, whose shape depends on temperature. As the temperature decreases through a critical value T_c, two equivalent minima appear, and the order parameter must choose one of them, thus spontanously breaking the symmetry.

Landau observes that, in a phase transition, the system becomes more "orderly" as it cools through the transition temperature. He proposes to measure the orderliness by a field ϕ called the order parameter. Modeled after the magnetization density, it should be zero above the critical temperature, and non-zero below it. In a state of thermodynamic equilibrium without external field, it should become uniform in space, with a value such as to minimize a certain potential energy.

The potential energy depends on temperature, and is assumed to have the shape illustrated in Fig. 18.4. At high temperatures it has a single minimum at $\phi = 0$, as shown in Fig. 18.4(a), and there is no symmetry breaking. When the temperature goes below a critical value, the potential energy develops two equivalent minima, as shown in Fig. 18.4(b). Now ϕ must choose one of the minima, and becomes non-zero. In doing so, it spontaneously breaks the symmetry of the potential energy.

The equilibrium value of ϕ reproduces the behavior of the magnetization shown in Fig. 18.1. It varies continuously with the temperature, but its slope jumps at the critical temperature.

Potential energy

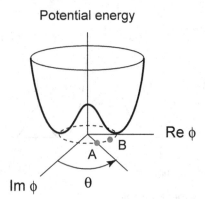

Fig. 18.5 Symmetry-breaking potential energy for system with complex order parameter. The potential is a figure of rotation, shaped like a wine bottle, here plotted over the complex plane of the order parameter. The lowest state is any point on the dotted circle along the trough at the bottom of the wine bottle, like A or B. The Goldstone mode is an excited state in which the order parameter is not uniform in space, but runs around the dotted circle as one moves in space.

18.4. The Goldstone mode

Consider now an order parameter that is a complex number:

$$\phi = Re^{i\theta}$$

The potential energy is assumed to have a wine bottle shape, as shown in Fig. 18.5. It is independent of the phase angle θ, and thus has global gauge invariance.

The lowest state in this potential lies on the dotted circle shown in the figure. Every point on the circle is a candidate for the equilibrium state, but only one can be realized, say point A. Choosing this spontaneously breaks the global gauge invariance, because a particular phase angle θ is singled out.

In the equilibrium state, the field sits at A at all points of space. If we had chosen B, then the field would sit at B at all points of space. Now consider a excited state, in which the phase angle slides from A to B as we change position in space, say along the x axis. This motion

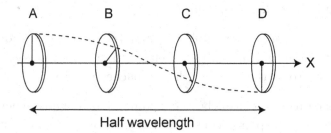

Fig. 18.6 Goldstone mode: the phase of the order parameter changes periodically as one advances in space, here along the x direction. The points A, B, C, D correspond to points on the dotted circle in Fig. 18.5. The energy of this mode of motion goes to zero as the wavelength goes to infinity.

Jeffrey Goldstone (1933–)

Fig. 18.7 Goldstone proves that spontaneous symmetry breaking means that the symmetry is expressed in a different manner, namely the existence of a excitation whose energy goes to zero as the wavelength goes to infinity. This is the called the Goldstone mode.

is illustrated in Fig. 18.6, and corresponds to an excitation. The potential energy is unchanged, but total energy is increased through the twist of the angle θ. It seems intuitively obvious that the energy increase can be made very small by making the wavelength of the motion very long.

Jeffrey Goldstone proves that

whenever a continuous global symmetry is broken spontaneously, there will appear an excited state whose energy approaches zero as the wavelength approaches infinity.

This excitation is called the "Goldstone mode", whose emergence
is an expression of spontaneous symmetry breaking. It is ubiquitous,
as the following table shows:

System	Broken symmetry	Goldstone mode
Magnet	Rotation	Spin wave
Solid	Translation	Phonon
Liquid helium	Global gauge invariance	Phonon
Superconductor	Local gauge invariance	N/A

The Goldstone mode is absent when the symmetry being broken
is local gauge invariance instead of global gauge invariance, and this
is what we take up next.

18.5. Superconductivity: the photon gets mass

The superconductor and liquid helium are both described by a complex order parameter ϕ, as discussed in the last section. It corresponds to the wave function of a Bose–Einstein condensate. The
difference is that for the superconductor the order parameter carries electric charge, and is coupled to the electromagnetic field. This
extends the global gauge invariance to a local one.

With local gauge invariance, there cannot be a Goldstone mode,
for the change of phase angle θ with position is a local gauge transformation, and has no effect on the system. Instead, the equation of
motion for the gauge field becomes

$$\nabla^2 \mathbf{A} + |\phi|^2 \, \mathbf{A} = 0 \,.$$

This means that the photon acquires mass $|\phi|$ in the superconducting
medium.

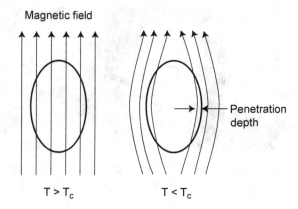

Fig. 18.8 Meissner effect: when a superconductor is cooled below the critical temperature, it expels magnetic fields from the interior, and the field can penetrate the body only to a finite depth. This means that inside the superconductor the photon has a mass equal to the inverse of the penetration depth. This is a manifestation of the spontaneous breaking of local gauge invariance.

This is manifested experimentally as the Meissner effect, namely, a magnetic field cannot exist inside the superconductor, but can only penetrate it to a finite depth. The penetration depth corresponds to the inverse mass of the photon. This is illustrated in Fig. 18.8. The electromagnetic field inside the superconductor is no longer transverse, as it is outside, but acquires a longitudinal component. This components comes from the degree of freedom that would have been the Goldstone boson.

In summary,

> when local gauge invariance is spontaneous broken, the gauge field "eats" the Goldstone boson and "gets fat", i.e., acquires mass.

18.6. Historical note

Landau conceived the order parameter around 1947. In 1950, he and Vitaly Ginzburg used it to construct a phenomenological theory

Vitaly L. Ginzburg   Philip W. Anderson   Peter Higgs
(1916–2009)   (1923–)   (1929–)

Fig. 18.9 Ginzburg and Landau proposed a phenemological model of superconductivity based on a complex order parameter. Anderson clarified the role of spontaneous breaking of local gauge invariance. Higgs applied these ideas to the Standard Model of particle interactions.

of superconductivity, before the discovery of the microscopic BCS (Bardeen, Cooper and Schrieffer) theory of 1957.

Philip Anderson, and later Y. Nambu, elucidated the Meissner effect in superconductivity, and superfluidity in liquid helium, in terms of spontaneous symmetry breaking. C. N. Yang advanced the idea of "off-diagonal long-range order" to supplement the order parameter in quantum systems.

In 1957, the microscopic BCS theory of superconductivity explains the phenomenon as a kind of Bose–Einstein condensation of composite bosons made up of a pair of electrons of opposite spin — the Cooper pair. The Ginzburg–Landau order parameter in fact represents the condensate wave function of the Cooper pairs. The BCS theory furnishes a dynamical description of the spontaneous breaking of local gauge invariance.

Goldstone's work was inspired by the Ginzburg–Landau model of superconductivity and the understanding in terms of symmetry breaking. In a relativistic context, excitations are characterized by its mass, and the Goldstone mode corresponds to a massless Goldstone

John Bardeen   Leon N. Cooper   J. Robert Schrieffer
(1908–1991)   (1930–)   (1931–)

Fig. 18.10 Creators of the BCS theory of superconductivity, which furnishes a dynamical description of spontaneous breaking of local gauge invariance. The phenomenlogical order parameter of Ginzburg and Landau emerges as the condensate wave function of paired electrons — the Cooper pairs.

boson. In the Standard Model, the closest thing to a Goldstone boson is the pion (Chap. 20).

Peter Higgs introduced the complex order parameter in the Standard Model that causes spontaneous breaking of local gauge symmetry, and generates mass for particles and gauge photons. The order parameter here is called the "Higgs field", and the Ginzburg–Landau way to generate photon mass is called the "Higgs mechanism". Experimentally, the vacuum value of the Higgs field is

$$|\phi| = 174 \text{ GeV.}$$

19

Quark Confinement

19.1. Monopole confinement

Quarks and gluons are not observed in isolation, but deduced as constituents of hadrons. Since quarks and gluons carry color charge, while hadrons do not, this suggests the rule that only "colorless" states can exist. This principle is called "color confinement", or "quark confinement".

Color charge generates color flux lines, just as electric charge generates electric flux lines, and these cost energy per unit length. An explanation of quark confinement is that a color charge will seek out neutralizing partners to form a bound state, in order to have the flux lines contained in a microscopic volume, thus minimizing energy.

Quark confinement has not been proven in QCD, because of mathematical complexities. It is widely accepted as plausible, however, because of an analogous phenomenon that we do understand — monopole confinement in a superconductor.

As we mentioned in the last chapter in connection with the Meissner effect, a superconducting medium tends to expel magnetic flux to lower the energy.

If we place an imaginary monopole and antimonopole into a superconductor, the magnetic flux created by them will be squeezed into a thin tube connected the two poles, as shown in Fig. 19.1(a). The agent enforcing this configuration is an induced solenoidal supercurrent, which arises as the response of Cooper pairs to the magnetic flux imposed. The flux tube is equivalent to a string of magnetic

Magnetic flux   Induced supercurrent

(a)

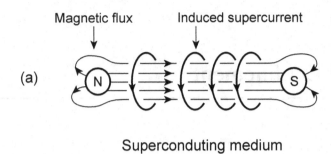

Superconduting medium

(b)

Fig. 19.1 Monopole confinement in a superconductor: the medium responds to the presence of magnetic flux by genreating a supercurrent that confines it in a tube. (a) A monopole and antimonopole will form a bound state tied by a flux tube. (b) The flux tube is equivalent to a string of dipoles.

dipoles, as illustrated in Fig. 19.1(b). It exerts a tension that draws the two poles together, and the size of the bound state is determined by a balance between this attraction and a short-range repulsion preventing the poles from overlapping each other.

If we imagine placing a single monopole in the medium, the magnetic flux would be contained in a flux tube leading from the monopole all the way to the surface of the superconductor. The energy cost will be enormous, since the flux tube would be of macroscopic length. The tube will try to contract, and in so doing will pull the monopole out to within a penetration depth from the surface. Thus, even if monopoles exist, we will find never find an isolated one inside a superconductor.

19.2. Electric flux tube

To understand quark confinement by analogy, we liken the vacuum to a superconducting medium, and the quark to a monopole. The difference is that magnetic field has to be replaced by color electric field. The picture is complicated by the fact that quarks have eight color charges, which are non-commutative. This means that their effects do not add in a simple way.

In Abelian electromagnetism, there is duality between electric and magnetic fields. The free Maxwell's equations are invariant under replacement of the electric field by the magnetic field, and the magnetic field by the negative of the electric field. The asymmetry between electric and magnetic phenomena arises solely from the fact that magnetic monopoles have not been observed.

In non-Abelian QCD, on the other hand, there is an absolute distinction between electric and magnetic fields. Electric charges are generators of the gauge group. There is no duality because of the non-linearity of the theory.

By analogy with the magnetic Meissner effect, which arises from a condensation of electron pairs, there should be a condensation of color-magnetic monopoles in the QCD vacuum to give rise to the electric Meissner effect. As a consequence, color electric flux lines will be squeezed into a thin flux tube, with finite energy per unit length. An imagined isolated quark in the vacuum would be trailed by a flux tube of cosmic dimension, and will be disposed of somehow, perhaps whooshed out to the "end" of the universe.

A quark–antiquark pair can be in a bound state with finite energy, tied together by an electric flux tube of microscopic length, as illustrated in Fig. 19.2(a). Here, group properties of the color-electric charge become relevant.

A quark belongs to the 3-dimensional fundamental representation of color $SU(3)$, denoted as $\mathbf{3}$. An antiquark belongs to the dual representation $\bar{\mathbf{3}}$.

A meson is a quark–antiquark system $q\bar{q}$, which can exist in a number of possible "irreducible" representations, with dimensionality determined by group theory. The group arithmetic for the $q\bar{q}$ system is

$$\mathbf{3} \times \bar{\mathbf{3}} = \mathbf{1} + \mathbf{8} \,,$$

where $\mathbf{1}$ corresponds to the colorless singlet state. A baryon is com-

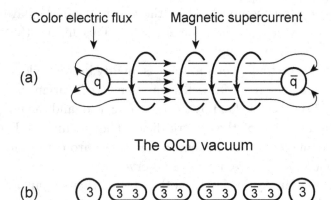

Fig. 19.2 Color confinement in QCD: The vacuum responds to color-electric flux by generating magnetic currents that squeeze the flux distribution into a tube. (a) A meson consists of a quark–antiquark pair tied by such a flux tube. (b) A quark is a **3**, in the language of group representation, and an antiquark is a **$\bar{3}$**. The flux tube is color-equivalent to a string of quark–antiquark pairs.

posed of qqq, and the group arithmetic reads

$$\mathbf{3} \times \mathbf{3} \times \mathbf{3} = \mathbf{1} + \mathbf{8} + \mathbf{8} + \mathbf{10} ,$$

and the singlet state **1** corresponds to a physical baryon.

From the standpoint of color, qqq is like $q\bar{q}$. This is because qq has a **$\bar{3}$** component, as indicated by the group arithmetic

$$\mathbf{3} \times \mathbf{3} = \bar{\mathbf{3}} + \mathbf{6} .$$

We then combine **$\bar{3}$** (in qq) with **3** (from the remaining q) to get **1**.

In Fig. 19.2(b) the flux tube is represented as a sequence of **$\bar{3}$-3** dipoles, from the point of view of color structure. This shows the tube "transports" color from one end to the other, where it gets neutralized. When the flux tube breaks, the hadron becomes two other hadrons instead of isolated quarks. This is like a bar magnetic breaking into two other bar magnets instead of two monopoles.

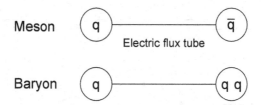

Fig. 19.3   A hadron can be modeled as quark and antiquark connected by a string representating a color-electric flux tube. This works for baryons composed of three quarks, because two quarks are color-equivalent to an antiquark.

19.3. The QCD string

We simplify the flux tube into a string, as in Fig. 19.3, and depict both meson ($q\bar{q}$) and baryon (qqq) as two color objects tied together by a string. As explained earlier, qq is color-equivalent to $\bar{q}$.

Bound states of the heavy quarks c, b, t can be described through a non-relativistic Schrödinger equation, with the string described by a linear potential energy. Such a model yields the calculated spectra of charmonium and bottomonium shown in Fig. 16.7.

The light quarks u, d, s can be treated as massless. Hadrons composed from them are modeled as rotating strings with massless quarks capping the ends, which move tangentially at the speed of light. The total energy of the rotating string gives the hadron's mass M, which turns out to be related to the spin J by

$$J = \alpha' M^2,$$

where $2\pi\alpha'$ is the inverse of the string tension. As shown in Fig. 19.4, this prediction agrees very well with experiments. The experimental value

$$\alpha' \approx 1 \ (\text{GeV})^{-2}$$

leads to the fascinating result:

$$\text{String tension} \approx 16 \text{ tons.}$$

Here is a truly awesome source of power, if only you could set a quark free.

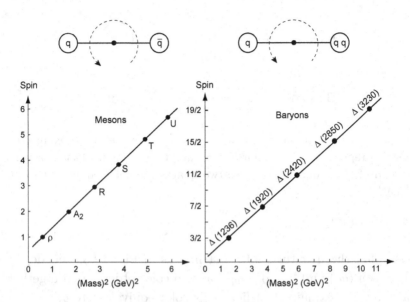

Fig. 19.4 In the string model, hadron are modeled by relativistically rotating strings capped with massless quarks at both ends. The model predicts that spin is proportional to mass squared, which is brilliantly confirmed by experimental data shown here. Observed particles are labeled by name. The slope of $1(\text{GeV})^{-2}$ corresponds to a string tension of 16 tons.

The QCD string realizes Tullio Regge's theory of recurrent resonances, which inspired Geoffrey Chew's bootstrap theory that morphed into modern string theory, "the theory of everything". (Chap. 15.) The plots in Fig. 19.4 were known as "Chew–Frautschi plots".

19.4. Asymptotic freedom

Quark confinement, like superconductivity, is a feature of a low-temperature phase of the system. We expect a phase transition at some high temperature that will liberate the quarks. An indication of this is the fact that the strong interaction weakens at high energies. The renormalized gauge coupling constant in QCD "runs" towards zero in the limit of infinite energy. This phenomenon is called

Frank Wilczek   David J. Gross   David Politzer
(1951–)   (1941–)   (1949–)

Fig. 19.5 Discoverers of asymptotic freedom.

"asymptotic freedom", first noticed by David Gross, Frank Wilzcek, and David Politzer.

In QED, the running charge has the opposite behavior. It is small at low energies, corresponding to $\alpha^{-1} \approx 137$, and grows indefinitely with increasing energy. As explained in Sec. 12.5, this is because we probe the electron to a smaller distance at higher energies. As we penetrate the cloud of induced charges that dresses the electron, we see more of the bare charge.

QCD exhibits the opposite behavior, because the gauge photons are themselves charged. The charge center of a bare quark shifts whenever a gluon is absorbed or emitted. As a result, the bare charge is smeared over a neighborhood, and there is no point charge at the center of a dressed quark. The charge contained in a volume element goes to zero when the size of the element shrinks to zero. This is a statement of asymptotic freedom.

In Fig. 19.6, we compare the charge distributions of a dressed electron and dressed quark, together with the Feynman diagrams describing the dressing due to vacuum polarization.

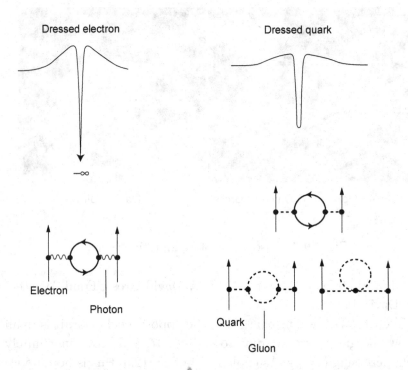

Fig. 19.6 Asymptotic freedom: the color charge distribution of a dressed quark (right) is smooth, in contradistinction to the charge distribution of a dressed electron (left), which contains a point charge. Thus, the color charge in a small volume goes to zero with its size. This is because photons are neutral, whereas gluons are charged. The charge center of a quark shifts upon emission or absorption of a gluon. Dressing of the particles are described by Feynman diagrams shown at the bottom. Asymptotic freedom arises from the two lower diagrams with gluon self-interaction.

20

Hanging Threads of Silk

The Chinese Empress took ill, and the court physician was summoned. He needed to take pulse for diagnosis, but protocol forbade him to touch the Empress. So the physician ordered silk threads tied to the finger tips of the Empress. While the hanging threads danced ever so subtly and delicately to the Empress' pulse, the physician observed, and rendered his diagnosis.

There are delicate issues in the Standard Model:

- Chiral invariance.
- PCAC.
- Triangle anomaly.

These issues predate quarks. Although we see them with more clarity in the quark picture, we still do not know their true origin. And so they remain dangling, like the Empress' silk threads.

20.1. Mass

Mass in the Standard Model is a dynamical property, not an intrinsic attribute as in Newtonian mechanics. The basic particles, leptons and quarks, are massless as required by gauge invariance. They acquire effective mass by a spontaneous breaking of gauge invariance via the vacuum value of the Higgs field.

What we observe in the laboratory, however, are not quarks, but hadrons made up of quarks. The masses of hadrons composed of the light quarks u and d have no direct relation to the quark masses, and

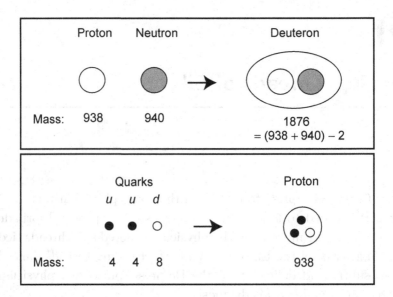

Fig. 20.1 Upper panel: the proton and neutron bind into the deuteron, whose mass is less than the sum of proton and neutron masses. Lower panel: light quarks bind into the proton, whose mass is sixty times the mass of its constituents. (All masses in MeV.)

have little to do with the Higgs field. Let us review the evidence for this.

In atoms and nuclei, the mass of a bound state is smaller than the sum of masses of the constituents, and the difference is called binding energy. For example, the mass of the hydrogen atom is smaller the sum of the proton mass and electron mass by 13 eV (energy equivalent). The proton and neutron have respective masses 938 and 940 MeV, but their bound state, the deuteron, has a mass of 1876 MeV, showing a binding energy of 2 MeV.

For a hadron composed of the light quarks, the situation is very different. The theoretical quark masses are respectively 4 and 6 MeV, but the proton made up of u u d has a mass of 938 MeV. The quark masses are negligible compared to this.

The contrast between deuteron binding and proton binding is illustrated in Fig. 20.1.

Chirality

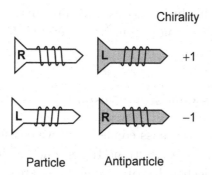

Particle   Antiparticle

Fig. 20.2 Chirality is +1 if a massless particle is right-handed, and antiparticle left-handed. Chirality is −1 if the reverse. Massive particles cannot have definite chirality because they cannot have intrinsic handedness.

The proton mass, in fact, emerges from a new kind of spontaneous symmetry breaking — that of *chiral invariance*.

20.2. Chirality

Chirality is a property of massless right- or left-handed neutrinos (R or L):

$$\text{Chirality} = \begin{cases} +1 & \text{If particle is R, antiparticle is L} \\ -1 & \text{If particle is L, antiparticle is R} \end{cases}.$$

This definition is illustrated in Fig. 20.2.

A massive spinning particle cannot have definite chirality, because its handedness is not permanent; the spin relative to the motion can be reversed by bringing the particle to rest and starting it off in the opposite direction. A massless particle cannot be stopped because it is always moving at the speed of light, and therefore a massless spinning particle has permanent handedness.

A chiral transformation changes the quantum phase of the wave function by an amount proportional to chirality. It is a global gauge transformation for which chirality is the "charge".

A theory that is invariant under chiral transformations can only have intrinsically massless particles. Put another way, to insure that particles have zero intrinsic mass, we require chiral invariance.

Then, an effective mass can be generated by spontaneously breaking chiral invariance.

20.3. The pion as Goldstone boson

To see where the proton mass comes from, we consider an idealized QCD with only light quarks with no Higgs coupling. The electroweak coupling is also neglected. In this model, there are only the quarks u and d coupled to color gauge fields. They are intrinsically massless, and therefore the theory is invariant under chiral transformations. This is called the *chiral limit*.

Because of the smallness of the quark masses, the chiral limit should be a good approximation to the hadronic world pertaining to light quarks. In this limit, there is perfect symmetry between left and right. All hadrons composed of light quarks should come in chiral-conjugate pairs with exactly the same mass. For example, the proton should have a partner with the same mass but opposite intrinsic parity. This is obviously not the case in the real world.

Yoichiro Nambu, and independently Zhou Guangzhao (Chou Kuang-Chao), concluded from such an analysis that the chiral symmetry is spontaneously broken. The symmetry is a global gauge symmetry and is not gauged. Its spontaneous breaking is therefore manifested through the existence of a massless Goldstone boson. They identify the pion as the Goldstone boson.

In the real world, the quark masses act as a small perturbation to this picture, and the Goldstone boson would acquire a small mass. This explains why the pion has such a small mass:

$$\frac{m_\pi}{m_p} \approx 0.15 \,.$$

Spontaneously symmetry breaking usually has a dynamical cause. For example, the breaking of local gauge invariance in superconductivity is due to a condensation of Cooper pairs, which arise from an attractive interaction between electrons induced by lattice

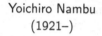

Yoichiro Nambu   Zhou Guangzhao
   (1921–)   (1929–)

Fig. 20.3 Why the pion has such a small mass: it is the Goldstone boson emerging from the spontaneous breaking of chiral symmetry in an idealized world. In the real world, chiral symmetry is only approximate, and the pion mass is close to zero.

vibrations. The chiral symmetry breaking must have a dynamical origin in QCD, but we have not yet understood the mechanism.

20.4. PCAC

Subtle phenomena arise from the near masslessness of the pion.

Chirality is the charge that generates chiral transformations, just as electric charge generates electromagnetic gauge transformations. Chiral invariance implies the existence of a conserved chiral current j_5, which is the analog of the conserved electromagnetic current j:

$$\partial \cdot j = 0,$$
$$\partial \cdot j_5 = 0, \text{ in chiral limit } (m_\pi \to 0).$$

In the real world, the chiral current is said to be "partially conserved".

The spatial components of j form a vector, meaning that its mirror image has an opposite direction. The spatial components of j_5

Marvin L. Goldberger   Sam Bard Treiman
(1922–)   (1925–1999)

Fig. 20.4 PCAC: the pion field is the source of the chiral current.

however, does not change sign upon reflection. It is called an "axial vector" instead of a vector.[1]

Marvin Goldberger (1922–) and Sam Treiman (1925–1999) suggested a way to calculate transition amplitudes involving pions, based on the partial conservation of the chiral current. The scheme is known as PCAC (partially conserved axial current), the second most awkward name for a theory.[2]

The idea is that

the pion field is the source of the chiral current.

More specifically, $\partial \cdot j_5$ is identified as an effective pion field. This makes it possible to obtain transition matrix elements between states containing pions. The scheme is successful in many practical applications, but its significance lies in the one glaring failure that opened the door to a deep mystery.

---

[1]The subscript 5 comes from the structure of quark currents. The charge current j^μ is built from the Dirac matrices γ^μ. The chiral current j_5^μ is built from $\gamma_5\gamma^\mu$, where γ_5 is the "fifth" Dirac matrix $\gamma^0\gamma^1\gamma^2\gamma^3$.

[2]The most awkward name for a theory is "Einstein's theory of the A and B coefficients", which deals with the spontaneous emission of photons by atoms.

The charged pions decay via the weak interaction into leptons:

$$\pi^{\pm} \rightarrow \mu^{\pm} + \begin{cases} \nu' \\ \bar{\nu}' \end{cases} \qquad (\text{lifetime} = 2.6 \times 10^{-8} \text{ s}).$$

The lifetime determines the phenomenological coefficient in PCAC amplitudes, which lead to an industry called "soft pion physics".

One then attempts to address the neutral pion, which is observed to decay into two photons, with a much shorter lifetime:

$$\pi^0 \rightarrow \gamma + \gamma \qquad (\text{lifetime} = 8.4 \times 10^{-17} \text{ s}).$$

The comparative shortness of the lifetime is due to the fact that this decay proceeds via the electromagnetic instead of weak interaction.

A routine calculation using PCAC fails; it predicts a much longer lifetime that goes to infinity in the chiral limit. That is, in the chiral limit the neutral pion would be stable.

The failure indicates that there must be additional contributions to the neutral pion decay, which has no effect on charged pion decay, and persists in the chiral limit. This is the so-called "triangle anomaly".

20.5. The triangle anomaly

A natural process is by definition "natural". It can seem "anomalous" only to the small bias mind, but that's us. We like to organize the world into neat packages that we understand, and any deviation from that would be considered anomalous.

The triangle anomaly occurs in an attempt to calculate the neutral pion decay, not through PCAC, but directly via Feynman diagrams. It is also called the ABJ anomaly after its discoverers Stephen L. Adler, John S. Bell, and Roman Jackiw.

The Feynman diagram for neutral pion decay is shown in Fig. 20.6. The pion dissociates virtually into $u\,\bar{u}$ and $d\,\bar{d}$ at an interaction vertex that invokes the chiral current j_5. The pair then annihilates into two photons via two interaction vertices involving the charge current j. The diagram is characterized by a triangular virtual quark loop.

Fig. 20.5 Clockwise: Stephen L. Adler (1939–), John S. Bell (1928–1990), and Roman Jackiw (1939–) discovered a profound mystery — the ABJ anomaly, also known as the triangle anomaly. (Picture of Bell taken in 1962 in Olympic National Park, WA, by the author.)

In the chiral limit with massless quarks, we naively expect the diagram to vanish. This is because j_5 flips the chirality of the circulating quark, but j does not. So when the quark goes around the loop it becomes orthogonal to the original state. This would imply that the neutral pion cannot decay in the chiral limit, which was the conclusion from PCAC.

A careful calculation, however, yields a non-zero result, and quantitatively explains the observed lifetime of the neutral pion. The revelation is that, instead of the conservation property $\partial \cdot j_5 = 0$

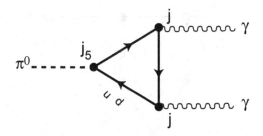

Fig. 20.6 Triangle anomaly: the Feynman diagram here does not vanish in the limit of massless quarks, contrary to naive expectation. This enables the neutral pion to decay. The decay rate is tripled when quark colors are taken into account, and agreement with experiments provides evidence for color.

in the chiral limit, one has instead

$$\partial \cdot j_5 = \frac{\alpha}{2\pi} \tilde{F} \cdot F,$$

where $\alpha \approx 1/137$ is the fine structure constant. The quantity $\tilde{F} \cdot F = \tilde{F}^{\mu\nu} F_{\mu\nu}$ is equal to $-4\mathbf{B} \cdot \mathbf{E}$ in terms of the magnetic field $\mathbf{B}$ and electric field $\mathbf{E}$. This beautiful but enigmatic result is the *triangle anomaly*.

The anomaly gives the correct lifetime for the neutral pion, provided we remember to triple each quark contribution because of color. Thus, the anomaly provides evidence for the existence of color as a bonus.

20.6. Lepton-quark family structure

With the triangle anomaly, however, there looms potential disaster.

The anomaly contributes to electron–neutrino scattering through the Feynman diagram shown in Fig. 20.7. The circulating fermion loop represents quarks and leptons in the first electroweak family (Chap. 17):

$$\left\{ \begin{matrix} \nu & e \\ \vec{u} & \vec{d} \end{matrix} \right\}.$$

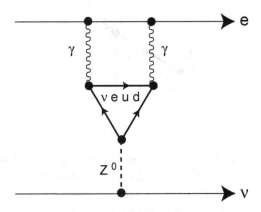

Fig. 20.7 The triangle subdiagram renders this diagram infinite, for any one fermion circulating around the triagular loop. This infinity cannot be renormalized away, and would be a real disaster. However, it cancels when contributions from the fermion family $\{\vec{u}\ \vec{d}\ e\ \nu\}$ are added up. This is a reason for the family structure.

The diagram with a particular fermion in the triangular loop has an ultraviolet catastrophe. Although the loop itself is finite, the high-frequency virtual photons attached to it cause a divergence. Unlike the divergence in QED, however, this one is a real disaster. It cannot be renormalized away because of wrong behavior under scale change.

However, we must add up the contributions from all the fermions in the family, and the result depends on the coupling constants determined by group structure. When that is done, lo and behold, the individual contributions cancel. Instead of infinity, one gets zero. Color copies of the quarks are needed for the cancellation. Thus, again, the anomaly gives evidence for color.

The cancellation of anomalies suggests that the lepton–quark family is an inseparable multi-component particle. Isolating any one component calls up uncontrollable high-frequency modes that are not really there. We do not know where the family comes from.

The anomaly represents a kind of spontaneous symmetry breaking because a current that ought to be conserved turns out not to be. However, it is different from the kind of breaking describable through an order parameter like the Higgs field. The chiral symmetry here is broken only in quantum theory. It remains unbroken in the classical theory. The phenomenon must have something to do with fluctuations involving quantum excitations.[3]

20.7. Waiting for closure

The Standard Model is like a volume opened in the middle, and we do not know how far we are from the end. In the continuing series that is physics, we have gone through several volumes:

• Classical physics,
• Quantum mechanics,
• Quantum electrodynamics.

We can close the book after each stage; the story seemed finished. A remarkable feature in all of the above stages is that the only building blocks of importance were proton, neutron, electron, and photon.

In contrast, the Standard Model is not a closed book. This is clear by glancing at the spectrum of "elementary particles" then and now, as shown in Fig. 20.8. The spectrum of the standard model looks like some sort of "periodic table", representing a "chemistry" of an underlying system of simpler structure.

There are many open issues:

• What is the Higgs field really made of?
• What causes quark confinement?
• What causes chiral symmetry breaking? Who perturbs it with light-quark masses?
• What is the origin of the triangle anomaly?
• What dictates the structure of the lepton-quark family?

---

[3]The anomaly appears to be related to topological excitations, for it involves the topological density $\tilde{F} \cdot F$. See K. Huang. *Quarks, Leptons, and Gauge Fields*, 2nd edn. (World Scientific, Singapore, 1992) Secs. 12.5, 12.6; K. Huang, *Quantum Field Theory: From Operators to Path Integrals* (Wiley, New York, 1998) Sec. 19.8.

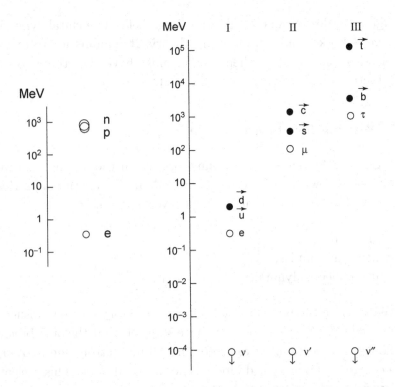

Fig. 20.8 Fundamental particles then and now, displayed in logarithmic mass scale. Left: nucleon, electron, photon (not shown) were all that is needed to build classical physics, quantum mechanics, and QED. Right: spectrum of the Standard Model has three families each containing eight particles. In addition, there are twelve gauge photons and a Higgs field of uncertain origin (not shown). Arrows on quarks indicate three colors. Neutrino masses are upper limits.

- Who ordered the three families? Why do their centers-of-mass increase exponentially?
- Why are neutrino masses so extremely small?

A larger question:

- Why is the gauge group $SU(3) \times SU(2) \times SU(1)$?

An even larger question:

- Why the gauge principle?

In "grand unified" theories, the gauge group is subsumed in a simple group like $SU(10)$. In "supersymmetric" theories, there is boson–fermion symmetry. But, aside from the lack of an experimental basis, they make the "chemistry" more complicated instead of simpler.

The one unifying theme of the Standard Model is the gauge principle; but that may not be the last word. Mathematical models indicate that it is possible for a gauge field to emerge from a simpler underlying theory.[4]

Only experiments can show us the way.

---

[4]An example is the "$O(3)$ non-linear σ-model", which describes a vector whose tip is constrained to move on a sphere. Another example is the "t-j model" of high-temperature superconductivity, which is a lattice model of very simple structure. In both cases, a gauge field arises as a way to solve a constraint.

The World in a Grain of Sand

21.1. A matter of scale

The poet William Blake (1757–1827) wrote:

> To see a world in a grain of sand ...

What world would we see? It depends on resolving power. As the sharpness of our perception changes, so changes the face of the world. At any particular resolution, we see an aspect that could be very different from that on other scales.

Equally important, the world looks the same over a vast stretch of length scales. That is, it appears to be self-similar over a range of magnification. That gives us time to linger, savor, and philosophize. Our world view can seem to be so compelling that we assume it to be the only possible truth. When increasing magnification brings us to the end of a self-similar range, with revelation of new structures, we need a "scientific revolution".

Take the image of a woodcut in Fig. 21.1, copied in a scanner with a particular resolution. Over a wide range of relatively low magnifications, it can be perceived through the human eye, registered in the brain, and evoke emotions. The "theorists" who make sense of this world are art critics, historians, and philosophers.

As we increase the magnification, there comes a point when we reach the resolving power of the scanner, as shown in the last panel of Fig. 21.1. What we see is the cutoff imposed by instrumentation. To interpret this as art would be absurd (or extraordinary creativity).

Fig. 21.1 Within a wide range of magnifications, a woodcut scanned at a particular resolution appears as art. However, there is a cutoff point, determined by the resolution of the scanner, beyond which the image reflects the temperament of the scanner rather than the artist.

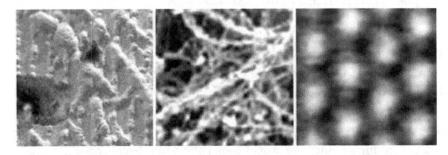

Fig. 21.2 The woodcut of the last figure is examined under microscopes of increasing power, progressing from the optical microscope, electron microscope, to tunneling electron microscope. The world revealed is no longer that of art, but material science. In the final scene we can make out individual atoms, and reach the threshold of the quantum world.

To go beyond the cutoff imposed by the scanner, we have to examine the original woodcut under microscopes of increasing magnifying power, as shown in Fig. 21.2. We see the texture of the paper on which the woodcut was imprinted, and leave the world of art for that of material science. Finally, through the tunneling electron microscope, we reach the end of this range of scales, and make out individual atoms. Beyond this lies the quantum world.

The character of the perceived world can change drastically when we go to a different scale. It seems futile to extrapolate what we know at one scale to the world that lies beneath. At the level of woodcut art, the repertory of our fantasy is not likely to include quantum mechanics. Newton must have understood this, when he said

> I seem to have been only like a boy playing on the seashore, and diverting myself in now and then finding a smoother pebble or a prettier shell than ordinary, whilst the great ocean of truth lay all undiscovered before me.

21.2. Renormalization

Our introduction to renormalization started with the ultraviolet catastrophe in QED — the divergence of integrals due to high-frequency modes. To obtain finite numbers to work with, theorists had to cut off the integrals, reluctantly. They were pleasantly surprised to discover that the cutoff could be "renormalized" away, i.e. absorbed into the observed mass and the charge of the electron. This enabled them to calculate physical quantities that agree with experiment to great precision.

Any theory we create cannot be expected to be valid at all length scales. It must fail at some small scale, and be replaced by a more correct theory. In the cutoff theory, modes of motion involving small length scales are not explicitly taken into account. This gives us a coarse-grained picture of the system.

A small length scale is equivalent to a high-frequency scale, since energy is proportional to frequency in quantum mechanics. In a particular situation, the relevant scale corresponds to the resolving power of the measuring instruments we use.

The cutoff is a parameter with dimension, and it sets a length scale. A renormalizable theory is one in which there is no intrinsic length scale other than the cutoff. After renormalization, the cutoff is no longer visible; but information about the scale resides in the renormalized parameters, and they change with a change of scale.

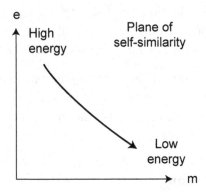

Fig. 21.3 Under a scale change, QCD remains self-similar, but with different values of the electron's renormalized charge e and mass m. The arrow points along the direction of increasing length scale, or decreasing energy scale.

As illustrated in Fig. 21.3, under a change of scale, the electron's renormalized charge and mass change, but the theory remains self-similar. That is, the theory is unchanged except for the values of these parameters.

Renormalizability is not just a property of QED, but of all successful theories in physics. The important point is that

a renormalizable theory describes phenomena at a particular length scale, in terms of parameters that can be measured at that scale.

For example, we can explain the everyday world using thermodynamics, without invoking atoms. Properties such as specific heat and thermal conductivity, which really originate from atomic structure, can be treated as empirical parameters. At a smaller length scale atoms appear, and they can be described by treating the nucleus as a point. Similarly, at the scale of nuclear structure we do not need quarks, and so forth.

Renormalizability is a closure property that makes physics possible. We would not be able to understand the world, if we had to understand every minute detail all at once.

Murray Gell-Mann   Francis E. Low
(1929–)   (1921–2007)

Fig. 21.4 Murray Gell-Mann and Francis Low pioneered the idea that the renormalized charge is a running coupling constant, i.e. it changes with the length scale.

21.3. The running coupling

In renormalized QED, once we obtain the electron charge at a particular scale from experiments, we can calculate its value at another scale. The calculation is particularly simple when the length scale is sufficiently short (or energy scale sufficiently high) so that we can neglect the electron's mass, which amounts to 0.5 MeV. In this manner, Murray Gell-Mann and Francis Low showed in 1954 that the charge increases logarithmically with energy.

Gell-Mann and Low suggested the physical picture that the bare charge is being screened by charges induced in the Dirac sea (Chap. 12). Thus, the effective charge seen by a probe depends on the distance from the bare charge, and for this reason is called a "running" coupling constant.

Years later, David Gross and Frank Wilczek, and independently David Politzer, found an opposite behavior in QCD, namely that the running color charge decreases logarithmically at high energies, approaching zero in the limit of infinite energy. Called "asymptotic freedom", this is due to fact that the gauge photons in QCD carry charge. When a color charge emits or absorbs virtual gauge photons, the original point bare charge is smeared out (Chap. 19).

Nikolay N. Bogolyubov	Curtis G. Callan	Kurt Symanzik
(1909–1992)	(1942–)	(1923–1983)

Fig. 21.5 Nikolay Bogolyubov introduced the idea of a renormalization group, and described how the coupling "runs" under scale change. An explicit equation was independently formulated by Curtis Callan and Kurt Symanzik.

Nikolai Bogolyubov viewed the scale change as a group operation, and proposed the idea of a *renormalization group* (RG) in 1967. Under scale change, the renormalized parameters trace out a trajectory called the *RG trajectory*. An equation for this trajectory was obtained by Curtis Callan and Kurt Symanzik independently in 1970.

In general, a renormalizable theory is characterized by an RG trajectory in a space spanned by a definite and fixed number of parameters. This is schematically depicted in Fig. 21.6, where the arrow on the trajectory points along the direction of increasing coarse-graining (decreasing energy-cutoff). The notches mark equal intervals of change in the cutoff. The "bare" theory we initially write down corresponds to some very high energy scale, and is denoted by an open circle. The renormalized theory corresponds to what we observe at a lower energy, and is marked by the solid circle.

In Fig. 21.7, we show the qualitative behavior of the running coupling constants in the Standard Model. The curve marked "strong" shows the QCD gauge coupling constant, which decreases with energy, exhibiting asymptotic freedom. The weak and the electromagnetic couplings both increase with energy. These curves appear to converge to a common value at the ultrahigh energy of 10^{16}–10^{20} GeV. It is tantalizing to think that a simpler theory

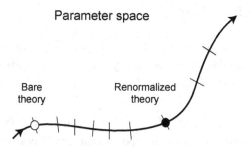

Fig. 21.6 RG (renormaliztion group) trajectory describing the running couplings (renormalized parameters) in a renormalizable theory. The arrows points along direction of coarse-graining. The theory was designed at a small distance scale marked by the open circle. We observe it at a larger distance scale marked by the solid circle.

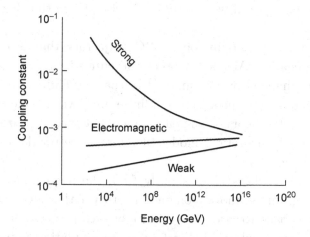

Fig. 21.7 Qualitative plot of the running coupling constants for strong, electromagnetic, and weak interactions in the Standard Model. They appear to converge at an energy of 10^{16}–10^{20} GeV, suggesting that a unification of interactions may happen at that scale.

presides at that scale. Suggestions so far center around "grand unified theories" that are gauge theories with different gauge groups, multitudes of Higgs fields, and seem to create more complications than they explain. A true unification may require radically different ideas.

21.4. Fixed point: theoretical model

During the early development of renormalization, theorists thought that the cutoff can be disposed of by "sending it to infinity" while holding renormalized parameters fixed. Actually, this cannot be done in QED. (More later.)

We cannot change the cutoff by mere declaration, because it is the only scale parameter in a renormalizable theory. It corresponds to the scale of the bare theory on an RG trajectory. To change that initial scale, we have to move the theory to another point on the trajectory. The question, therefore, is whether there exists a trajectory such that the bare theory can be placed at a point corresponding to infinite cutoff.

The answer is yes, if we can find a fixed point on that trajectory. Let us explain.

A fixed point is a point on an RG trajectory that is invariant under scale change. At such a point the cutoff is infinite, because it cannot be changed by any amount of coarse-graining. If we want the cutoff in our bare theory to be infinite, we have to tune the bare parameters of the theory in such a manner that theory is situated at a fixed point. A theoretical model therefore corresponds to a fixed point in parameter space.

If we displace the system slightly from the fixed point, it will, upon coarse-graining, move along a trajectory, either away from the fixed point or back towards it. The rate of such motion goes to zero at the fixed point, which is therefore an accumulation point of the tick marks of Fig. 21.6.

When the system goes away upon coarse-graining, the fixed point is seen by the system as an ultraviolet (UV) fixed point, since it lies at a higher energy scale. The reverse corresponds to an infrared (IR) fixed point.

21.5. UV fixed point: QCD

An example of a UV fixed point is that in QCD, which exhibits asymptotic freedom. That means at high energies the coupling runs

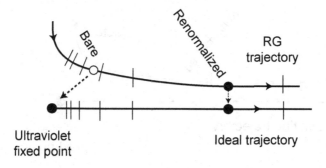

Fig. 21.8 QCD is governed by an ultraviolet fixed point located at zero coupling. The ideal theory corresponds to the lower RG trajectory, with the bare theory at the fixed point. The upper trajectory corresponds a theory we set up initially. We can tune the parameters of the theory so that we approach the ideal trajectory.

to zero, which is therefore a UV fixed point. This is schematically illustrated in Fig. 21.8. The ideal QCD is governed by an ultraviolet fixed point located at zero coupling. When we set up the theory with a finite cutoff, the bare system lies on a neighboring trajectory shown above the ideal trajectory, at a point marked by the open circle. To make the cutoff of the bare theory go to infinity, we tune the bare parameters in such a manner that the trajectory approaches the ideal one, and the bare system approaches the fixed point.

In a theory with asymptotic freedom, such as QCD, we can fulfill the wish of "sending the cutoff to infinity" while holding renormalized parameters fixed.

21.6. IR fixed point: QED

In contrast to QCD, the running coupling in QED increases indefinitely as the energy increases. The fixed point occurs in the opposite direction, in the low-energy limit. The theory is governed by an IR fixed point at zero coupling.

The situation is very different from that in QCD, for the RG trajectory that contains an IR fixed point is not a trajectory in the

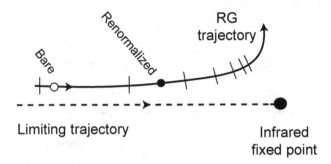

Fig. 21.9 QED is governed by an infrared fixed point located at zero coupling. The RG trajectory going into the fixed point, shown dotted, is a singular line along which the cutoff is infinite, and the coupling is zero everywhere. The physical theory lies on a trajectory close to it, with non-zero renormalized coupling taken from experiments.

proper sense, but a limiting curve for trajectories. Under coarse-graining, the cutoff can only decrease, because the energy scale is lowered. But since the fixed point has infinite cutoff, all points on the trajectory going into an IR fixed point must have infinite cutoff. The renormalized coupling is zero along the entire trajectory.

The bare theory cannot be located on the limiting trajectory because the cutoff is infinite along it. It has to be on a neighboring trajectory, as illustrated in Fig. 21.9. When this trajectory is made to approach the limiting trajectory the tick marks on the approaching trajectory will be spaced further and further from each other, until they are infinitely apart in the limit.

If we take the QED trajectory to be exactly the dotted limiting curve in Fig. 21.9, then we would have to conclude that the charge of the electron is zero. Known as "triviality", this property was pointed out by Landau, as we reported in Chap. 15. This "triviality" applies to any coupling that is not asymptotically free, such as that in the Higgs sector of the Standard Model.

By taking the electron charge to be given by experiments, instead of zero, we are placing the theory not on the limiting trajectory, but on some trajectory close to it. The cutoff has an unknown finite value, which we need not know. This trajectory represents Dyson's "jerry-built" structure (Chap. 12). But, it agrees with experiments to one part in a trillion.

21.7. Crossover: scientific revolution

As long as a theory is self-similar, you can never get out of that theory through scaling. As the scale changes, you go along an RG trajectory confined to a fixed parameter space. In reality, the world we perceive can change drastically under a change of scale. This means the "true" trajectory must be able to break out of the confines of the old parameter space, and veer into new dimensions.

The true RG trajectory should be close to a theoretical trajectory in the neighborhood of a fixed point. In this neighborhood the fixed point is a good approximation to reality. As long as we are close enough to the fixed point, a scale change advances the system only a small distance along the trajectory, for that is what a fixed point means. The system also appears to be self-similar (renormalizable), because the theoretical trajectory has this property.

As we keep changing the scale, however, there comes a point when the true trajectory veers away, going into a new dimension unknown to the old theory. Freed from the old fixed point, the system advances rapidly on its trajectory, only to be lured and captured by the next fixed point.

Like a cruise ship, the true theory has made many ports of call while traveling up the energy scale:

Classical physics → Quantum mechanics → QED → Yang–Mills.

In renormalization terms going from one port to the next is a *crossover*; in sociological terms it is a "scientific revolution".

Crossovers are unknown to the renormalization theory of Dyson, Gell-Mann and Low, Bogolyubov, and Callan and Symanzik, because their theory remains self-similar at all scales. To enable the trajectory to make a crossover, we have to adopt a larger view of renormalization, and this is the subject of the next chapter.

22

In the Space of All Possible Theories

22.1. The physics is in the cutoff

Renormalization has snatched victory from the jaws of the ultra-violet catastrophe in QED. It has managed to bury the cutoff in renormalization parameters, and magically made it disappear from sight. Despite impressive agreement with experiments, the process will remain magical until we understand its physical basis. For that we have to realize that the cutoff is a physical parameter as emphasized by Kenneth Wilson.

Theories in physics deal with phenomena observed within certain ranges of length scales. Newtonian mechanics describes motion we see in the everyday world, quantum mechanics takes over at a scale measured in angstroms (10^{-8} cm), and QED goes down to 10^{-13} cm, etc. Any theory of our design has a limit of applicability, even though we may not know what it is, and wish that there were none.

We can specify a theory by giving the Lagrangian at the smallest scale of applicability. All modes of motion with higher frequency than a cutoff Λ_0 are ignored. This defines the "bare theory".

At a lower frequency scale, we adopt a coarse-grained picture, by effectively lower the cutoff from Λ_0 to a smaller value Λ. This should be done not by expunging the modes between the two cutoffs, but by "hiding" them in such a way that the theory appears to have a new cutoff Λ_1 without any change in substance. The process is illustrated in Fig. 22.2.

The result of coarse-graining would be an effective Lagrangian with a new cutoff. The new Lagrangian should describe exactly the

Kenneth G. Wilson (1936–)

Fig. 22.1   How to navigate in the space of Lagrangians.

same system as before, and only the appearance changes. Under repeated coarse-graining, we should generate a sequence of effective Lagrangians tracing out the RG trajectory in the space of Lagrangians, as depicted symbolically in Fig. 22.3. Thus, the trajectory describes the appearance of the system when examined under varying resolving power.

22.2. The RG trajectory

Kenneth Wilson implements the coarse-graining as follows.

The Lagrangian, as we recall, is the kinetic energy minus the potential energy:

Lagrangian = Kinetic energy − Potential energy.

We can choose the theory by specifying the nature of the basic field, and the form of the kinetic and potential energies as functions of the field. The field as a function of time, with given initial and final

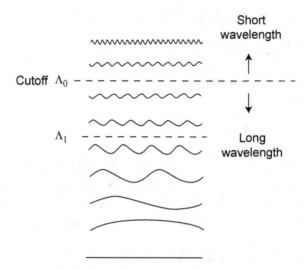

Fig. 22.2   In the bare theory, all modes with frequencies higher than the cutoff Λ_0 are ignored. Coarse-graining lowers the effective cutoff to Λ_1, by "hiding" the modes between Λ_0 and Λ_1 without changing the theory.

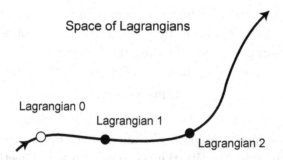

Fig. 22.3   Under coarse-graining, the effective Lagrangian moves on the RG trajectory. The basic system being described remains unchanged; only its appearance changes.

configurations, constitutes a "history", or Feynman path. The cutoff is introduced by limiting acceptable paths to those involving frequencies less than some value. We can do this by composing a path by a superposition of frequency components, much like vibrating modes of a string, and cut off the spectrum after a frequency Λ_0:

$$\text{Frequency components of path: } f_1, f_2, \ldots, \quad \Big|_{\text{Cutoff}\Lambda 0} .$$

This defines the bare Lagrangian, labeled with a subscript 0. It gives the bare action

$$A_0(P) = \int_P dt \, \text{Lagrangian}_0 .$$

The integration over time t is carried out along a particular path P, and the action depends on the path. A complete description of the system is given by the Feynman path integral, which gives the transition probability amplitude between given endpoints:

$$\text{Amplitude} = \sum_P \exp \frac{i}{\hbar} A_0(P)$$

$$= \sum_{f_1} \sum_{f_2} \cdots \sum_{f_{\Lambda 0}} \exp \frac{i}{\hbar} A_0(f_1, f_2, \ldots, f_{\Lambda 0}) ,$$

where the sum extends over all cutoff paths with specified endpoints. We can sum over paths by summing over each frequency-component independently, as indicated in the second line of the above formula.

Coarse-graining is done by hiding the frequency components between Λ_0 and Λ_1, so as to lower the effective cutoff to Λ_1:

$$f_1, f_2, \ldots, \quad \Big|_{\Lambda 1} \overset{\text{Hide these modes}}{\cdots\cdots} \Big|_{\Lambda 0} .$$

The "hiding" is done as follows. We sum over the frequencies between the two cutoffs, and identify the result with a new effective action A_1:

$$\sum_{f_{\Lambda 1}} \cdots \sum_{f_{\Lambda 0}} \exp \frac{i}{\hbar} A_0(f_1, f_2, \ldots, f_{\Lambda 0}) = \exp \frac{i}{\hbar} A_1(f_1, f_2, \ldots, f_{\Lambda 1}) .$$

The new action now depends only on frequencies below Λ_1, and the original Feynman amplitude now takes the form

$$\text{Amplitude} = \sum_{f_1} \sum_{f_2} \cdots \sum_{f_{\Lambda 1}} \exp \frac{i}{\hbar} A_1(f_1, f_2, \ldots, f_{\Lambda 1}) .$$

The new effective action corresponds to a new effective Lagrangian, which involves paths with lower frequencies than Λ_1. The frequencies between Λ_0 and Λ_1 have been "absorbed"; their effects are felt only through the changed Lagrangian. By repeating this manner, we generate the RG trajectory

$$\text{Lagrangian}_0 \rightarrow \text{Lagrangian}_1 \rightarrow \text{Lagrangian}_2 \rightarrow \cdots .$$

22.3. The space of Lagrangians

The true RG trajectory moves in the space of Lagrangians as the scale changes. There is no reason why the effective theory should be self-similar. However, there are fixed points in this space, in general. The trajectory will slow down as it approaches a fixed point, and in its neighborhood the trajectory may remain in some restricted subspace, and thus appear to represent a self-similar system. However, it will veer away from the fixed point after a while, and resume its journey, until it approaches another fixed point.

As illustration, we depict the approach of the true trajectory to a self-similar one representing QED, starting from high frequencies. Couplings not relevant to QED should become "irrelevant", i.e. tend to zero as the trajectory approaches the self-similar plane spanned by the mass and charge of the electron. These irrelevant parameters include the Weinberg–Salam couplings from the unification with the weak interaction, and the yet unobserved ones signifying deviations from QED.

How large is the space of Lagrangians? As large as is necessary to accommodate all Lagrangian that satisfy canonical requirements. A new Lagrangian may emerge with a different potential energy, requiring a reorganization of the old field variable. This will give the theory a new look.

We can argue that there is always a solution for the effective Lagrangian, given a sufficiently general space. Perhaps the most general system we can imagine is one built from binary integers. Coarse-graining is just a reorganization of the rules governing their relations.

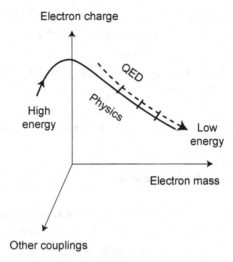

Fig. 22.4 The trajectory of physics almost coincides with that of QED for a range of length scales, but deviates from it at both ends. In the short-distance, or high-frequency end, the trajectory comes in from other dimensions corresponding to unification with the weak interactions, and to deviations from QED not yet discovered. The low-energy end goes to classical physics.

In this sense, the space available to the RG trajectory is the space of all possible theories.

22.4. Of time and temperature

One of the remarkable properties of quantum field theory is that it formally reduces to statistical mechanics, when time becomes pure imaginary. Specifically, the time t is related to the absolute temperature T (in energy units) through

$$t = \frac{i\hbar}{T}.$$

Under this identification, the Feynman amplitude (with specially chosen endpoints) maps into the partition function of statistical mechanics.

Leo P. Kadanoff   Michael E. Fisher
(1937–)   (1931–)

Fig. 22.5 Kadanoff described spin systems under coarse-graining. Fisher explored phases near a critical point.

Wilson's original renormalization scheme is formulated in terms of the partition function. This is not surprising in view of the fact that the existence of a cutoff is natural in the theory of matter, because the physics requires it. In the crystal lattice of a solid, for example, the lattice spacing is a natural cutoff. It relieves us from worrying about quarks while calculating thermal conductivity.

The main achievement of the Wilson method lies in the description of critical phenomena in the theory of phase transitions, in particular the calculation of critical exponents. Its contribution to renormalization in quantum theory is a physical understanding of its basis.

A precursor of the Wilson scheme was Leo Kadanoff's method of coarse-graining in a spin lattice. He discovered that spins interacting via nearest-neighbor interactions become block spins that acquire next to next-nearest neighbor interactions, etc. This generates an RG trajectory.

Michael Fisher contributed to the calculation of critical exponents. He also clarified properties of RG trajectories, including fixed points and crossovers.

Why there should be a connection between time and temperature remains a mystery.

Fig. 22.6   Qu Yuan (340–278 BC).

22.5. *Tian Wen* (天问)

The ancient Chinese poet Qu Yuan (屈原), unhappily exiled from the court of the King of Chu, roamed the landscape aimlessly, and finally drowned himself in a river. To this day, people stage dragon boat races on the anniversary of his death, ostensibly to save him. They also prepare sweet-rice offerings wrapped in palm to throw in the river, to keep him from starvation. In his wanderings, Qu Yuan put brush to cliff sides and walls of caves, demanding answers to philosophical and historical questions. The collection came to be known as *Tian Wen* (Ask Heaven). The opening lines, in particular, continue to baffle scientists today:

遂古之初,	At the primordial beginning,
谁传道之?	Who was the reporter?
上下未形,	Before the universe took shape,
何由考之?	How could one measure it?

Nothing existed before the Big Bang: no space-time, no physical law. All that would have to spring up at the instant of the Big Bang, which could only take place at the fixed point of nothingness.[1] But the universe could not have been created exactly at a fixed point, for that would mean that no scale change is possible. It will have to be displaced infinitesimally from the null fixed point, either along a trajectory connected to the fixed point, or one passing nearby. The act of creation, therefore, consists of choosing a direction to kick the world out.

Trajectories could emanate from the null fixed point in an infinity of possible directions. Since there was nothing to begin with, there can be no rules governing their formation except logic. Thus, any theory that can be mathematically formulated is possible.[2] Some possibilities are represented schematically in Fig. 22.7:

- Trajectories that go into the fixed point upon coarse-graining define trivial theories.
- Trajectories going away from the fixed point define non-trivial, asymptotically free theories.
- Passing trajectories, such as T in Fig. 22.7, are dense in the neighborhood. The world could be placed on one such trajectory infinitesimally close to the null fixed point, but it is ultimately controlled by some other fixed point.

---

[1]Since the null set must be contained in any set, the only guaranteed existence is non-existence.

[2]In the simplest example of a scalar field theory, there are directions that lead to asymtotically free theories, contrary to a prevailing belief that only gauge theories can be asymptotically free. The false belief was based on the self-imposed constraint that the potential be a polynomial in the field. [K. Halpern and K. Huang, *Phys. Rev.* **53**, 3252 (1996); K. Huang, *Quantum Field Theory: From Operators to Path Integrals* (Wiley, New York, 1998) Chap. 17.] For lack of a better term, these are called "Halpern–Huang directions".

Fixed point of nothingness

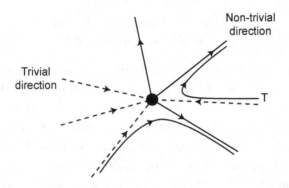

Fig. 22.7   Creation took place at the fixed point of nothingness. Trajectories emanating from the fixed point can be trivial or non-trivial, depending on whether it goes into or away from the fixed point upon coarse-graining. There are also passing trajectories (T) that come arbitrary close to the fixed point. The universe could be placed on any of these trajectories, perhaps at random.

Nothing seems to dictate placement; so it must have been chosen at random. That would mean that the physics was chosen at random from all possibilities. The systems placed on an IR trajectory will have no interaction; those placed on a UV trajectory will exhibit asymptotic freedom; and those on a passing trajectory can have any behavior logically permissible.

Are we just one among an infinitude of universes being spit out at random? Are we the fortunate inheritor of a universe that happens to "make sense"?

22.6. *Tian Wen* updated

- Could big bangs be happening continually?
- Could universes be born on random trajectories — and so sample all possible mathematical structures?
- They could not interact with us. Could they?

Epilogue: Beauty is Truth

A fixed point is a structure of pure mathematics, a thing of beauty:

> a beauty cold and austere, like that of sculpture, without appeal to any part of our weaker nature, without the gorgeous trappings of painting or music, yet sublimely pure, and capable of a stern perfection such as only the greatest art can show.[1]

Physics is truth. It sails down a trajectory in the space of Lagrangians, when the energy scale shrinks from that set by the Big Bang. It gets attracted to fixed points and lingers in their neighborhoods — as it must, by nature of fixed points. The journey thus proceeds from fixed point to fixed point, and only at these ports of call do we have the opportunity to observe and understand it. And at these times, beauty and truth become one.

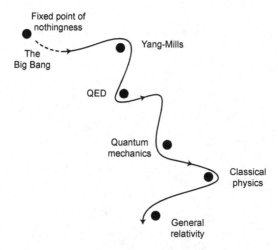

---

[1]Bertrand Russell, "Study of Mathematics" in *Mysticism and Logic* (Reprinted by Dover, New York, 2004).

Appendix

Nobel Prize in Physics

Many people mentioned in this book were honored by the Nobel Prize in Physics for their contributions. Rather than pointing that out in each case, we furnish a list of prize recipients, by year and in alphabetical order. Items referred to in the book are marked with an asterisk.

Annual listing

2006

The prize is being awarded jointly to:

JOHN C. MATHER and GEORGE C. SMOOT for their discovery of the blackbody form and anisotropy of the cosmic microwave background radiation.

2005

The prize is being awarded with one half to:

ROY J. GLAUBER for his contribution to the quantum theory of optical coherence,

and one half jointly to:

JOHN L. HALL and THEODOR W. HÄNSCH for their contributions to the development of laser-based precision spectroscopy, including the optical frequency comb technique.

2004

The prize is being awarded jointly to:

*DAVID J. GROSS, H. DAVID POLITZER and FRANK WILCZEK for the discovery of asymptotic freedom in the theory of the strong interaction.

2003

The prize is being awarded jointly to:

*ALEXEI A. ABRIKOSOV, VITALY L. GINZBURG and AN-THONY J. LEGGETT for pioneering contributions to the theory of superconductors and superfluids.

2002

The prize is being awarded with one half jointly to:

*RAYMOND DAVIS JR. and MASATOSHI KOSHIBA for pioneering contributions to astrophysics, in particular for the detection of cosmic neutrinos,

and the other half to:

RICCARDO GIACCONI for pioneering contributions to astrophysics, which have led to the discovery of cosmic X-ray sources.

2001

The prize is being awarded jointly to:

ERIC A. CORNELL, WOLFGANG KETTERLE and CARL E. WIEMAN for the achievement of Bose-Einstein condensation in dilute gases of alkali atoms, and for early fundamental studies of the properties of the condensates.

2000

The prize is being awarded with one half jointly to:

ZHORES I. ALFEROV and HERBERT KROEMER for developing semiconductor heterostructures used in high-speed- and opto-electronics, and one half to:

JACK ST. CLAIR KILBY for his part in the invention of the integrated circuit.

1999

The prize was awarded jointly to:

*GERARDUS 'T HOOFT, and MARTINUS J. G. VELTMAN for elucidating the quantum structure of electroweak interactions in physics.

1998

The prize was awarded jointly to:

ROBERT B. LAUGHLIN, HORST L. STORMER and DANIEL C. TSUI for their discovery of a new form of quantum fluid with fractionally charged excitations.

1997

The prize was awarded jointly to:

STEVEN CHU, CLAUDE COHEN-TANNOUDJI and WILLIAM D. PHILLIPS for development of methods to cool and trap atoms with laser light.

1996

The prize was awarded jointly to:

DAVID M. LEE, DOUGLAS D. OSHEROFF and ROBERT C. RICHARDSON for their discovery of superfluidity in helium-3.

1995

The prize was awarded for pioneering experimental contributions to lepton physics, with one half to:

*MARTIN L. PERL for the discovery of the tau lepton,

and the other half to:

*FREDERICK REINES for the detection of the neutrino.

1994

The prize was awarded for pioneering contributions to the development of neutron scattering techniques for studies of condensed matter to:

BERTRAM N. BROCKHOUSE for the development of neutron spectroscopy,
CLIFFORD G. SHULL for the development of the neutron diffraction technique.

1993

The prize was awarded jointly to:

RUSSELL A. HULSE and JOSEPH H. TAYLOR JR. for the discovery of a new type of pulsar, a discovery that has opened up new possibilities for the study of gravitation.

1992

GEORGES CHARPAK for his invention and development of particle detectors, in particular the multiwire proportional chamber.

1991

PIERRE-GILLES DE GENNES for discovering that methods developed for studying order phenomena in simple systems can be generalized to more complex forms of matter, in particular to liquid crystals and polymers.

1990

The prize was awarded jointly to:

*JEROME I. FRIEDMAN, HENRY W. KENDALL and RICHARD E. TAYLOR for their pioneering investigations concerning deep inelastic scattering of electrons on protons and bound neutrons, which have been of essential importance for the development of the quark model in particle physics.

1989

One half of the award was given to:

NORMAN F. RAMSEY for the invention of the separated oscillatory fields method and its use in the hydrogen maser and other atomic clocks,

and the other half jointly to:

HANS G. DEHMELT and WOLFGANG PAUL for the development of the ion trap technique.

1988

The prize was awarded jointly to:

*LEON M. LEDERMAN, MELVIN SCHWARTZ and JACK STEIN-BERGER for the neutrino beam method and the demonstration of the doublet structure of the leptons through the discovery of the muon neutrino.

1987

The prize was awarded jointly to:

J. GEORG BEDNORZ and K. ALEXANDER MÜLLER for their important breakthrough in the discovery of superconductivity in ceramic materials.

1986

The prize was awarded with one half to:

ERNST RUSKA for his fundamental work in electron optics, and for the design of the first electron microscope,
GERD BINNIG and HEINRICH ROHRER for their design of the scanning tunneling microscope.

1985

KLAUS VON KLITZING for the discovery of the quantized Hall effect.

1984

The prize was awarded jointly to:

*CARLO RUBBIA and SIMON VAN DER MEER for their decisive contributions to the large project, which led to the discovery of the field particles W and Z, communicators of weak interaction.

1983

The prize was divided equally between:

SUBRAMANYAN CHANDRASEKHAR for his theoretical studies of the physical processes of importance to the structure and evolution of the stars,
WILLIAM A. FOWLER for his theoretical and experimental studies of the nuclear reactions of importance in the formation of the chemical elements in the universe.

1982

*KENNETH G. WILSON for his theory for critical phenomena.

1981

The prize was awarded with one half jointly to:

NICOLAAS BLOEMBERGEN and ARTHUR L. SCHAWLOW for their contribution to the development of laser spectroscopy,

and the other half to:

KAI M. SIEGBAHN for his contribution to the development of high-resolution electron spectroscopy.

1980

The prize was divided equally between:

*JAMES W. CRONIN and VAL L. FITCH for the discovery of violations of fundamental symmetry principles in the decay of neutral K-mesons.

1979

The prize was divided equally between:

*SHELDON L. GLASHOW, ABDUS SALAM and STEVEN WEIN-BERG for their contributions to the theory of the unified weak and electromagnetic interaction between elementary particles, including, *inter alia*, the prediction of the weak neutral current.

1978

The prize was divided, with one half being awarded to:

PYOTR LEONIDOVICH KAPITSA for his basic inventions and discoveries in the area of low-temperature physics,

and the other half divided equally between:

ARNO A. PENZIAS and ROBERT W. WILSON for their discovery of cosmic microwave background radiation.

1977

The prize was divided equally between:

*PHILIP W. ANDERSON, SIR NEVILL F. MOTT and JOHN H. VAN VLECK for their fundamental theoretical investigations of the electronic structure of magnetic and disordered systems.

1976

The prize was divided equally between:

*BURTON RICHTER and SAMUEL C. C. TING for their pioneering work in the discovery of a heavy elementary particle of a new kind.

1975

The prize was awarded jointly to:

AAGE BOHR, BEN MOTTELSON and JAMES RAINWATER for the discovery of the connection between collective motion and particle motion in atomic nuclei and the development of the theory of the structure of the atomic nucleus based on this connection.

1974

The prize was awarded jointly to:

SIR MARTIN RYLE and ANTONY HEWISH for their pioneering research in radio astrophysics, Ryle for his observations and inventions, in particular of the aperture synthesis technique, and Hewish for his decisive role in the discovery of pulsars.

1973

The prize was divided, with one half being equally shared between:

LEO ESAKI and IVAR GIAEVER, for their experimental discoveries regarding tunneling phenomena in semiconductors and superconductors, respectively,

and the other half to:

BRIAN D. JOSEPHSON for his theoretical predictions of the properties of a supercurrent through a tunnel barrier, in particular those phenomena which are generally known as the Josephson effects.

1972

The prize was awarded jointly to:

*JOHN BARDEEN, LEON N. COOPER and J. ROBERT SCHRIEFFER for their jointly developed theory of superconductivity, usually called the BCS-theory.

1971

DENNIS GABOR for his invention and development of the holographic method.

1970

The prize was divided equally between:

HANNES ALFVÉN for fundamental work and discoveries in magneto-hydrodynamics with fruitful applications in different parts of plasma physics,

LOUIS NÉEL for fundamental work and discoveries concerning

antiferromagnetism and ferrimagnetism which have led to important applications in solid-state physics.

1969

*MURRAY GELL-MANN for his contributions and discoveries concerning the classification of elementary particles and their interactions.

1968

LUIS W. ALVAREZ for his decisive contributions to elementary particle physics, in particular the discovery of a large number of resonance states, made possible through his development of the technique of using hydrogen bubble chamber and data analysis.

1967

*HANS ALBRECHT BETHE for his contributions to the theory of nuclear reactions, especially his discoveries concerning the energy production in stars.

1966

ALFRED KASTLER for the discovery and development of optical methods for studying hertzian resonances in atoms.

1965

The prize was awarded jointly to:

*SIN-ITIRO TOMONAGA, JULIAN SCHWINGER and RICHARD P. FEYNMAN for their fundamental work in quantum electrodynamics, with deep-ploughing consequences for the physics of elementary particles.

1964

The prize was divided, with one half being awarded to:

CHARLES H. TOWNES,

and the other half jointly to:

NICOLAY GENNADIYEVICH BASOV and ALEKSANDR MIK-
HAILOVICH PROKHOROV for fundamental work in the field of
quantum electronics, which has led to the construction of oscillators
and amplifiers based on the maser-laser principle.

1963

The prize was divided, with one half being awarded to:

*EUGENE P. WIGNER for his contributions to the theory of the
atomic nucleus and the elementary particles, particularly through
the discovery and application of fundamental symmetry principles,

and the other half jointly to:

MARIA GOEPPERT-MAYER and J. HANS D. JENSEN for their
discoveries concerning nuclear shell structure.

1962

*LEV DAVIDOVICH LANDAU for his pioneering theories for con-
densed matter, especially liquid helium.

1961

The prize was divided equally between:

ROBERT HOFSTADTER for his pioneering studies of electron scattering in atomic nuclei and for his thereby achieved discoveries
concerning the stucture of the nucleons,
RUDOLF LUDWIG MÖSSBAUER for his researches concerning the
resonance absorption of gamma radiation and his discovery in this
connection of the effect which bears his name.

1960

DONALD A. GLASER for the invention of the bubble chamber.

1959

The prize was awarded jointly to:

EMILIO GINO SEGRÈ and OWEN CHAMBERLAIN for their
discovery of the antiproton.

1958

The prize was awarded jointly to:

PAVEL ALEKSEYEVICH CHERENKOV, IL'JA MIKHAILOVICH FRANK and IGOR YEVGENYEVICH TAMM for the discovery and interpretation of the Cherenkov effect.

1957

The prize was awarded jointly to:

*CHEN NING YANG and TSUNG-DAO LEE for their penetrating investigation of the so-called parity laws which has led to important discoveries regarding the elementary particles.

1956

The prize was awarded jointly, with one third each, to:

WILLIAM SHOCKLEY, JOHN BARDEEN and WALTER HOUSER BRATTAIN for their researches on semiconductors and their discovery of the transistor effect.

1955

The prize was divided equally between:

*WILLIS EUGENE LAMB for his discoveries concerning the fine structure of the hydrogen spectrum,
*POLYKARP KUSCH for his precision determination of the magnetic moment of the electron.

1954

The prize was divided equally between:

*MAX BORN for his fundamental research in quantum mechanics, especially for his statistical interpretation of the wavefunction,
WALTHER BOTHE for the coincidence method and his discoveries made therewith.

1953

FRITS (FREDERIK) ZERNIKE for his demonstration of the phase contrast method, especially for his invention of the phase contrast microscope.

1952

The prize was awarded jointly to:

FELIX BLOCH and EDWARD MILLS PURCELL for their development of new methods for nuclear magnetic precision measurements and discoveries in connection therewith.

1951

The prize was awarded jointly to:

SIR JOHN DOUGLAS COCKCROFT and ERNEST THOMAS SINTON WALTON for their pioneer work on the transmutation of atomic nuclei by artificially accelerated atomic particles.

1950

*CECIL FRANK POWELL for his development of the photographic method of studying nuclear processes and his discoveries regarding mesons made with this method.

1949

*HIDEKI YUKAWA for his prediction of the existence of mesons on the basis of theoretical work on nuclear forces.

1948

LORD PATRICK MAYNARD STUART BLACKETT for his development of the Wilson cloud chamber method, and his discoveries therewith in the fields of nuclear physics and cosmic radiation.

1947

SIR EDWARD VICTOR APPLETON for his investigations of the physics of the upper atmosphere especially for the discovery of the so-called Appleton layer.

1946

PERCY WILLIAMS BRIDGMAN for the invention of an apparatus to produce extremely high pressures, and for the discoveries he made therewith in the field of high pressure physics.

1945

*WOLFGANG PAULI for the discovery of the Exclusion Principle, also called the Pauli Principle.

1944

*ISIDOR ISAAC RABI for his resonance method for recording the magnetic properties of atomic nuclei.

1943

OTTO STERN for his contribution to the development of the molecular ray method and his discovery of the magnetic moment of the proton.

1942–1940

The prize money was allocated to the Main Fund (1/3) and to the Special Fund (2/3) of this prize section.

1939

*ERNEST ORLANDO LAWRENCE for the invention and development of the cyclotron and for results obtained with it, especially with regard to artificial radioactive elements.

1938

*ENRICO FERMI for his demonstrations of the existence of new radioactive elements produced by neutron irradiation, and for his related discovery of nuclear reactions brought about by slow neutrons.

1937

The prize was awarded jointly to:

CLINTON JOSEPH DAVISSON and SIR GEORGE PAGET

THOMSON for their experimental discovery of the diffraction of electrons by crystals.

1936

The prize was divided equally between:

VICTOR FRANZ HESS for his discovery of cosmic radiation,
*CARL DAVID ANDERSON for his discovery of the positron.

1935

*SIR JAMES CHADWICK for the discovery of the neutron.

1934

The prize money was allocated to the Main Fund (1/3) and to the Special Fund (2/3) of this prize section.

1933

The prize was awarded jointly to:

*ERWIN SCHRÖDINGER and PAUL ADRIEN MAURICE DIRAC for the discovery of new productive forms of atomic theory.

1932

*WERNER HEISENBERG for the creation of quantum mechanics, the application of which has, *inter alia*, led to the discovery of the allotropic forms of hydrogen.

1931

The prize money was allocated to the Main Fund (1/3) and to the Special Fund (2/3) of this prize section.

1930

SIR CHANDRASEKHARA VENKATA RAMAN for his work on the scattering of light and for the discovery of the effect named after him.

1929

*PRINCE LOUIS-VICTOR DE BROGLIE for his discovery of the wave nature of electrons.

1928

SIR OWEN WILLANS RICHARDSON for his work on the thermionic phenomenon and especially for the discovery of the law named after him.

1927

The prize was divided equally between:

ARTHUR HOLLY COMPTON for his discovery of the effect named after him,
*CHARLES THOMSON REES WILSON for his method of making the paths of electrically charged particles visible by condensation of vapor.

1926

*JEAN BAPTISTE PERRIN for his work on the discontinuous structure of matter, and especially for his discovery of sedimentation equilibrium.

1925

The prize was awarded jointly to:

*JAMES FRANCK and GUSTAV HERTZ for their discovery of the laws governing the impact of an electron upon an atom.

1924

KARL MANNE GEORG SIEGBAHN for his discoveries and research in the field of X-ray spectroscopy.

1923

ROBERT ANDREWS MILLIKAN for his work on the elementary charge of electricity and on the photoelectric effect.

1922

*NIELS BOHR for his services in the investigation of the structure of atoms and of the radiation emanating from them.

1921

*ALBERT EINSTEIN for his services to Theoretical Physics, and especially for his discovery of the law of the photoelectric effect.

1920

CHARLES EDOUARD GUILLAUME in recognition of the service he has rendered to precision measurements in Physics by his discovery of anomalies in nickel steel alloys.

1919

JOHANNES STARK for his discovery of the Doppler effect in canal rays and the splitting of spectral lines in electric fields.

1918

*MAX KARL ERNST LUDWIG PLANCK in recognition of the services he rendered to the advancement of Physics by his discovery of energy quanta.

1917

CHARLES GLOVER BARKLA for his discovery of the characteristic Röntgen radiation of the elements.

1916

The prize money for 1916 was allocated to the Special Fund of this prize section.

1915

The prize was awarded jointly to:
SIR WILLIAM HENRY BRAGG and SIR WILLIAM LAWRENCE BRAGG for their services in the analysis of crystal structure by means of X-rays.

1914

MAX VON LAUE for his discovery of the diffraction of X-rays by crystals.

1913

HEIKE KAMERLINGH-ONNES for his investigations on the properties of matter at low temperatures which led, *inter alia,* to the production of liquid helium.

1912

NILS GUSTAF DALÉN for his invention of automatic regulators for use in conjunction with gas accumulators for illuminating lighthouses and buoys.

1911

WILHELM WIEN for his discoveries regarding the laws governing the radiation of heat.

1910

JOHANNES DIDERIK VAN DER WAALS for his work on the equation of state for gases and liquids.

1909

The prize was awarded jointly to:

GUGLIELMO MARCONI and CARL FERDINAND BRAUN in recognition of their contributions to the development of wireless telegraphy.

1908

GABRIEL LIPPMANN for his method of reproducing colors photographically based on the phenomenon of interference.

1907

*ALBERT ABRAHAM MICHELSON for his optical precision instruments and the spectroscopic and metrological investigations carried out with their aid.

1906

*SIR JOSEPH JOHN THOMSON in recognition of the great merits of his theoretical and experimental investigations on the conduction of electricity by gases.

1905

PHILIPP EDUARD ANTON LENARD for his work on cathode rays.

1904

LORD JOHN WILLIAM STRUTT RAYLEIGH for his investigations of the densities of the most important gases and for his discovery of argon in connection with these studies.

1903

The prize was divided, with one half being awarded to:

ANTOINE HENRI BECQUEREL in recognition of the extraordinary services he has rendered by his discovery of spontaneous radioactivity,

and the other half jointly to:

PIERRE CURIE and MARIE CURIE, née SKLODOWSKA in recognition of the extraordinary services they have rendered by their joint researches on the radiation phenomena discovered by Professor Henri Becquerel.

1902

The prize was awarded jointly to:

*HENDRIK ANTOON LORENTZ and PIETER ZEEMAN in recognition of the extraordinary service they rendered by their researches into the influence of magnetism upon radiation phenomena.

1901

WILHELM CONRAD RÖNTGEN in recognition of the extraordinary services he has rendered by the discovery of the remarkable rays subsequently named after him.

Alphabetical listing

Hänsch, Theodor W. 2005
*Heisenberg, Werner 1932
*Hertz, Gustav 1925
Hess, Victor Franz 1936
Hewish, Antony 1974
Hofstadter, Robert 1961
*Hooft, Gerardus 't 1999
Hulse, Russell A. 1993

Jensen, J. Hans D. 1963
Josephson, Brian D. 1973

Kamerlingh-Onnes, Heike 1913
Kapitsa, Pyotr Leonidovich 1978
Kastler, Alfred 1966
*Kendall, Henry W. 1990
Ketterle, Wolfgang 2001
Kilby, Jack S. 2000
Klitzing, Klaus Von 1985
*Koshiba, Masatoshi 2002
Kroemer, Herbert 2000
*Kusch, Polykarp 1955

*Lamb, Willis Eugene 1955
*Landau, Lev Davidovich 1962
Laue, Max Von 1914
Laughlin, Robert B. 1998
*Lawrence, Ernest Orlando 1939
*Lederman, Leon M. 1988
Lee, David M. 1996
*Lee, Tsung-Dao 1957
Leggett, Anthony J. 2003
Lenard, Philipp Eduard Anton 1905
Lippmann, Gabriel 1908
*Lorentz, Hendrik Antoon 1902

Marconi, Guglielmo 1909
*Meer, Simon Van Der 1984
Mather, John C. 2006
*Michelson, Albert Abraham 1907
Millikan, Robert Andrews 1923
Moessbauer, Rudolf Ludwig 1961
Mott, Sir Nevill F. 1977
Mottelson, Ben 1975
Muller, K. Alexander 1987

Neel, Louis 1970

Osheroff, Douglas D. 1996

Paul, Wolfgang 1989
*Pauli, Wolfgang 1945
Penzias, Arno A. 1978
*Perl, Martin L. 1995
Perrin, Jean Baptiste 1926
Phillips, William D. 1997
*Planck, Max Karl Ernst Ludwig 1918
*Politzer, H. David 2004
*Powell, Cecil Frank 1950
Prokhorov, Aleksandr
    Mikhailovich 1964
Purcell, Edward Mills 1952

*Rabi, Isidor Isaac 1944
Rainwater, James 1975
Raman, Sir Chandrasekhara
    Venkata 1930
Ramsey, Norman F. 1989
Rayleigh, Lord John William
    Strutt 1904
*Reines, Frederick 1995
Richardson, Robert C. 1996
Richardson, Sir Owen Willans 1928
*Richter, Burton 1976
Roentgen, Wilhelm Conrad 1901
Rohrer, Heinrich 1986
*Rubbia, Carlo 1984
Ruska, Ernst 1986
Ryle, Sir Martin 1974

*Salam, Abdus 1979
Schawlow, Arthur L. 1981
*Schrieffer, J. Robert 1972
*Schrödinger, Erwin 1933
*Schwartz, Melvin 1988
*Schwinger, Julian 1965
Segre, Emilio Gino 1959
Shockley, William 1956
Shull, Clifford G. 1994
Siegbahn, Kai M. 1981
Siegbahn, Karl Manne Georg 1924

Smoot, George C. 2006
Stark, Johannes 1919
*Steinberger, Jack 1988
Stern, Otto 1943
Störmer, Horst 1998
*Tamm, Igor Yevgenyevich 1958
Taylor, Joseph H. Jr. 1993
*Taylor, Richard E. 1990
Thomson, Sir George Paget 1937
*Thomson, Sir Joseph John 1906
*Ting, Samuel C. C. 1976
Tomonaga, Sin-Itiro 1965
Townes, Charles H. 1964
Tsui, Daniel C. 1998

Van Der Waals, Johannes Diderik 1910
Van Vleck, John H. 1977
*Veltman, Martinus J. G. 1999

Walton, Ernest Thomas Sinton 1951
*Weinberg, Steven 1979
Wieman, Carl E. 2001
Wien, Wilhelm 1911
*Wigner, Eugene P. 1963
*Wilczek, Frank 2004
*Wilson, Charles Thomson Rees 1927
*Wilson, Kenneth G. 1982
Wilson, Robert W. 1978

*Yang, Chen Ning 1957
*Yukawa, Hideki 1949

Zeeman, Pieter 1902
Zernike, Frits 1953

Name Index

Subject Index